Pumps and Hydraulic Rams

With Numerous Engravings and Diagrams

Paul N. Hasluck 1907

Reprinted by EnergyBook 2021 (Formally Onetoremember)

ISBN-13: 978-1468160260

ISBN-10: 1468160265

Edited by Richard Jemmett

Copyright © 2021 RW Jemmett

All rights reserved. Permission is granted to copy or reprint portions for any noncommercial use, except they may not be posted online without permission.

PREFACE

This Handbook contains, in a form convenient for everyday use, a number of articles on Pumps and Hydraulic Rams contributed by experts to Work and Building World—two weekly journals it is my fortune to edit.

Readers who may desire additional information respecting special details of the matters dealt with in this Handbook, or instructions on kindred subjects should address a question to Work, so that it may be answered in the columns of that journal.

P. N. HASLUCK.

La Belle Sauvage, London. April, 1907.

Contents List

SUCTION PUMPS AND LIFT PUMPS	4
MAKING SIMPLE SUCTION PUMPS	24
PUMP CUP LEATHERS	46
PUMP VALVES	52
RAM OR PLUNGER PUMPS	77
MAKING BUCKET AND PLUNGER PUMPS.	114
CONSTRUCTION OF PLUMBER'S FORCE PUMP	122
WOODEN PUMP	136
SMALL PUMPS FOR SPECIAL PURPOSES	145
CENTRIFUGAL PUMPS	155
AIRLIFT, MAMMOTH, AND PULSOMETER	162
HYDRAULIC RAMS	177
ABOUT THE PUBLISHER	203
DISCLAIMER	203

CHAPTER I

SUCTION PUMPS AND LIFT PUMPS

Years ago, before the majority of towns was supplied with water from companies' mains, there was an enormous number of pumps in everyday use for domestic purposes, and most of these were of either the suction type (jack pumps) or of that type known as the lift, force, or engine. Both types of pumps have still an extensive use in country districts, and more commonly come into the hands of the fitter and repairer than any other kind. Suction pumps and lift pumps do not differ in the manner in which they raise the water from the well into the barrel, but they do differ in the manner in which they expel the water from the barrel into the trough or supply pipe. The similarity and difference are clearly shown in the diagrams (Figures 1 and 2). Both have a pipe a descending into the well; in both cases, water is sucked into the barrel B on the upstroke of the piston C, the valve D opening automatically during the upstroke, and closing again when the piston reaches the top of its travel. But from this point the actions differ. In the suction pump (Figure 1), as the piston descends again, the piston valves E open, so forcing the water to flow through them; on the next upstroke the piston valves are closed, and the piston lifts up the water, which then flows out through the delivery pipe F; during this upstroke, of course, the piston has re-filled with water, and the cycle of events is then repeated. In the lift pump (Figure 2), the barrel

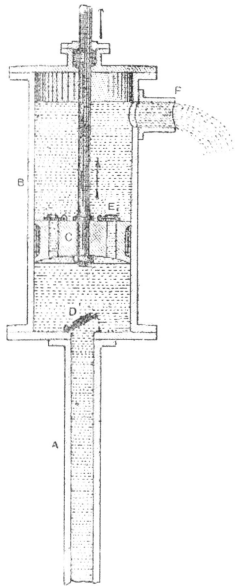

Figure 1 Suction Pump: Explanatory Diagram

having been filled as already described, the solid piston descends, the valve G in delivery pipe F opens, and the water is forced out to a height depending on

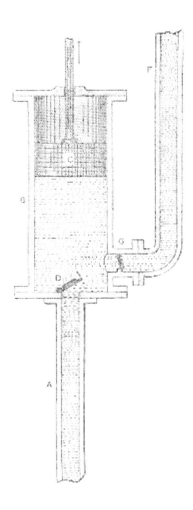

Figure 2 Single-acting Lift Pump: Explanatory Diagram.

the force with which the piston descends; then on the next upstroke the barrel re-fills.

A cast-lead suction or jack pump, as shown by Figure 3, is usually cast by the plumber himself. The action of this pump is as follows: —when the bucket a (Figure 3) is raised, a certain amount of air is drawn from the suction pipe B, and as the strokes are repeated all the air is exhausted, and a vacuum is formed. The pressure of the atmosphere, which is 14"

.7 lb. per square inch at sea level, exerts itself on the water in the well, and forces it up the suction pipe B, as there is no air in this pipe and consequently no weight to counteract the atmospheric pressure.

The water, however, will only rise to a certain height; this limit is reached when there is enough o f it in the pipe B (Figure 3) to balance the weight of the atmosphere.

To find the theoretical height to which the water will rise, divide the atmospheric pressure, in pounds per square inch (say, 14.7 lb.) by the weight of one square inch of water 1 ft. high. One sq. in. of water 1 ft. high = .434 lb.; 14.7 lb. ÷.434 lb. = 33.88, or 33 ft. 101/2 in. approximately.

In practice, the water would not rise more than 28 ft. or 29 ft. It is not wise to fix a suction or a lift pump more than 25 ft. from the surface of the water to the top of the bucket. Because of this limitation, a different class of pump is used for deep wells.

The jack pump's suction pipe b (Figure 3) is fixed to oak stages to support the pipe C, which forms the barrel in which the bucket works. D is the sucker, and E the handle. Figure 4 shows a section of the sucker, Figure 5 a section of the bucket.

The first thing to do when fixing a pump is to determine the size of the barrel, which is a most important matter. The head of water x. 4335 gives pounds pressure per square inch; thus 20 ft. head x •4335 = 8.67 lb. per square inch. Lever handles are

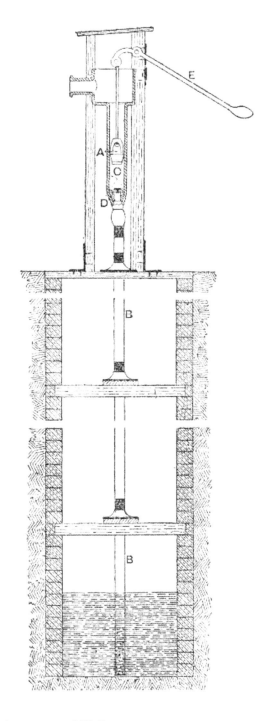

Figure 3 Lead Jack Pump and Well

usually set by makers at six to one, which means that for every 1 lb. exerted at one end of the lever 6 lb. is exerted at the other. The power of a man is reckoned at 20 lb. for continuous pumping. Multiply the power of the handle by the power of a man; thus, 20 lb. man x 6 lb. handle = 120 lb., which is the quantity the man can raise by this means. Divide the pounds raised by the man by the pressure, divide by .7854, and the square root of this will give the size of the barrel. Thus, 120 lb. $\div$ 18.67 lb. = 3.84 $\div$ .7854 =17, the square root of which is, say, 4 in., which is the size of the barrel. The deeper the well, the smaller the barrel must be, or the power must be correspondingly increased. When the size of the barrel has been found, the suction pipe can be fixed.

The size of the suction pipe should be half that of the barrel to which it is wiped, the lower end being bossed or plugged up and some holes drilled about a foot from the bottom to form a strainer to prevent dirt, etc., getting into the pump and choking the valves.

The sucker leather should be fixed on as shown at F (Figure 4). Oil-dressed leather of about the thickness of sole leather should be used. A lead clack is fastened to the leather to prevent the water getting out of the barrel, the clack weighting the leather down. To fix the sucker, bind some green hemp round it, and melt some tallow in a ladle and pour over it. Take the pump hook (Figure 6), screw the thread lightly into the lead clack, and lower it into the barrel. When the sucker is low enough to catch in the throat, unscrew the hook, and with the other end drive it home. Fix a leather valve and clack in the bucket in the same manner. Nail a piece of leather in the groove right round the outside of the bucket, as at g (Figure 5), the edges overlapping about 1 1/2 in., and being bevelled. Copper tacks are best for holding the leather, as they do not rust. The sucker and bucket should be soaked in water before the nailing is begun, as being of elm they are liable to split if nailed when dry. All leathers should be soaked and greased before being used.

The end of the bucket rod should be made red hot, and pushed through the top of the bucket and cooled quickly. The rod is held more tightly by burning the hole than by boring it. A washer and split key are used for holding the rod in place.

The handle should be so fixed that the bucket may be drawn up straight, not all on one side. The barrel ought to be cased in to prevent anything getting into it.

The most common cause of pumps becoming defective is due to the valves getting out of order, either by being worn out, or through things getting underneath the valves. The remedy is to fix new leathers. To take out the sucker, boiling water is poured down the pump in order to melt the tallow, or the outside of the pump is heated by a fire or with a blow lamp, the pump hook being inserted through the sucker, which is withdrawn by the barbed hook on the rod.

Sometimes the suction pipe leaks, and sucks in air. This fault can be easily detected by placing the ear on the pump while somebody is pumping, when, if the pipe is defective, a hissing noise will be heard.

Figure 4 Section of a Pump Sucker

Figure 5 Section of a Pump Bucket

Figure 6 Pump Hook

Another kind of jack pump (illustrated at Figure 7) is made of cast-iron, the barrel being bolted to a flange, and the bottom leather being between the flanges, making a water- and air-tight joint. The bucket is made of gun-metal, with a cup leather. The pump works on the same principle as that of the lead jack pump.

When the well is more than 25 ft. and less than 35 ft. deep, to save the expense of a lift pump a long barrel jack is used, the sucker or bottom valve being within the 25 ft., the barrel and the bucket making up the excess in length.

The pump used for lifting water to a greater

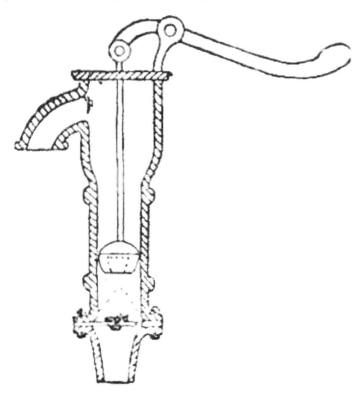

Figure 7 Iron Jack Pump

height than 25 ft. is called a lift-, force-, or engine- pump, as already explained. Note that the pump itself must not be fixed at a greater distance from the surface of the water than is the case with the ordinary suction pump. When a well is less than 25 feet deep the pump can be fixed above ground, but when the well is deeper the pump must be fixed in the well at any point below 25 ft. from the water.

Figure 8 shows a lift and force pump fixed above ground, H is the bucket, J the suction pipe, K the barrel, L the bottom valve, M the handle, N the top valve, O the sling and guide, and P the delivery pipe. The sling and guide are used for keeping the pump rod vertical. This kind of pump is usually made in brass, with a copper rod and an iron handle. The action is as follows: —When the bucket is worked and the air exhausted, forming a vacuum, the atmospheric pressure forces the water up the suction pipe, passing through the bottom valve L. When the bucket is lowered it closes valve L and forces the water in the barrel up through its own valve and through the top valve, which closes when the return stroke is made, thus holding a certain amount of water in the delivery pipe P. The strokes being repeated draw up more water, which rises through the top valve until the delivery pipe is filled and the water pours out. The valves are made of leather with lead clacks, and are bolted between flanges; their shape is shown at Figure 9.

When the leathers are to be removed, take out

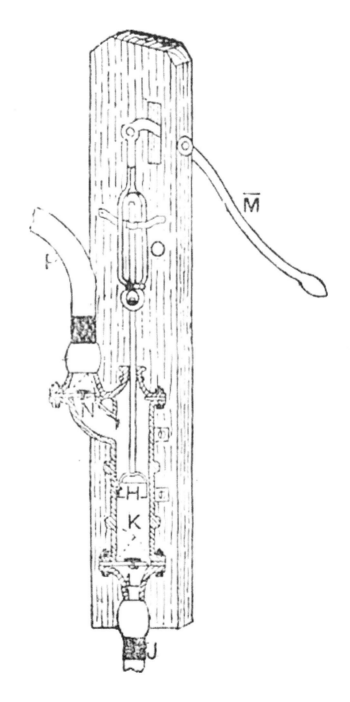

Figure 8 Lift Pump with Delivery Pipe connected to Top of Barrel

the set screws in the flanges, and remove the barrel and delivery pipe. A useful tool for cutting out leathers is the washer cutter illustrated at Figure 10. This can be regulated by the set screw, and consists of two knives worked like a pair of compasses. Another method of cutting out washers is to file one leg of a compass to a sharp edge and then use it in the ordinary manner; by both these methods leathers can be cut cleaner and truer than with a knife.

The lift pump, whose principle is illustrated by Figure 2, is single-acting; but double-acting lift pumps are also in use, their principle being made clear by Figure 11. In a single-acting lift pump, water is deli

 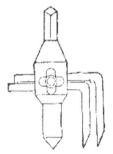

Figure 9 Pump Leather

Figure 10 Washer Cutter

vered on each alternate stroke; in a double-acting pump water is delivered on every stroke. In Figure 11, A indicates suction pipe in well; B pump barrel; C piston; D lower admission valve; and D' upper admission valve; F delivery pipe; G lower delivery valve; and G' upper delivery valve. Obviously as the piston on every stroke sucks in a supply it also expels a barrelful, the action of the automatic valves being such that one or other of the admission valves is always open and one or other of the delivery pipe valves is always open, too. The diagram offers a full explanation.

Sometimes, when the well is very deep, double and triple barrel pumps are used instead of single ones, but as the fixing of these is the same as for a single barrel, the latter only will be described.

It will be as well first to say something about working in deep wells. Never attempt to descend without first lowering down a lighted candle. If the candle burns, the well is safe to work in, as a light requires pure air in order that it may burn; if the air is foul, the candle will go out, and the well is therefore unsafe to work in till the foul air has been expelled. The foul air, when there is not a great deal of it, may be removed by slaking a pail of quicklime down below, the steam from which will force the foul air up. When there is a large quantity of foul air present, the only way is to pump fresh air down by means of a fan, or blow-forge as it is called. Whilst one man is working in a well, there should be another at the top in case of accident. If tools are sent down, they should be properly secured to prevent risk of injury to the workmen below, and also to prevent loss.

Figure 12 is a deep-well pump fixed in position: Q is the pump, R the suction pipe, S the delivery pipe, T the pump rod, U the handle, V the sling and guide, W the false bottom, X the stages, Y the main stage, and Z the air-vessel. The main stage, which is of oak 1 ft. square, should be built into the sides of the well about 20 ft. from the bottom; on no account should it be more than 25 ft. from the bottom.

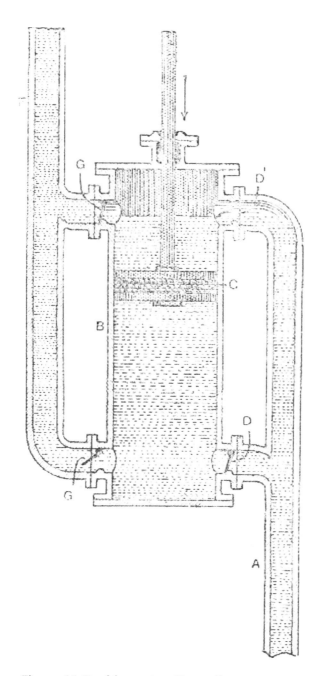

Figure 11 Double-acting Piston Pump:
Explanatory Diagram

The false bottom, made with oak boards, is fixed 2 ft. below the main stage, and forms a platform for standing on when working at the pump; it also prevents tools, etc., dropping into the water and becoming lost. The other stages are fixed 12 ft. apart, and should be of double 4-in. by 4-in. oak, with enough space left between them for the delivery pipe to pass through. Formerly the only way to descend a well was by means of a windlass and bucket, but now that stages are provided it is possible to go down from stage to stage by means of short ladders.

Having, by the method described earlier, found the size of the barrel, take out the screws from the bottom flange of the pump and remove the leather. The bottom part of the flange should be wiped to the suction pipe, which, like the delivery pipe, is half the size of the barrel. The bottom end of the suction pipe is bossed up and drilled as before. Instead of bossing it up, some plumbers prefer to drive a wood plug into the and of the pipe. Lower the suction pipe down the well, pass it through the main stage and false bottom, and bolt the flange to the former. Replace the bottom valve, after soaking and greasing the leather, and screw the pump to the flange. The flange of the delivery pipe can now be removed and the valve taken out, the object of removing these valves being to examine them; for if the pipes were wiped to the pump whilst the valves were in the latter, the heat would shrivel the leathers and render them useless. The pump can now be filled with water and tested to see if the bottom valve is sound, and the bucket rod can be pulled up and down a few times to see if it throws properly, and to see that it sucks in no air. In the latter case, it may be necessary to tighten the screws in the bottom flange. Cover the top flange to prevent anything falling into the barrel, and wipe the other half of the flange to the delivery pipe.

An air vessel should always be provided on a deep-well pump, as it saves much labour in working, keeps up a steady flow of water, and lengthens the life of the pump by reducing the vibration. Z (Figure 12) shows the air vessel fixed in place, Figure 13 being a section. It should be fixed as near the top valve as possible. Its action will be easily seen on reference to Figure 13. Before the pump is started, the air vessel and the pipes are full of air, and when the pump is worked this air is driven out of the delivery pipe, but is confined in the air vessel, which has no outlet. The air, in striving to get out, presses on the water in the same way as the atmosphere, and forces it up the pipe, thus helping the pump and making it work easier and more evenly. The air vessel is usually made of copper, and is wiped to the delivery pipe.

A piece of oak 11/2 in. thick, 8 in. wide, and long enough to span the width between the two halves of the stage, is dropped over the delivery pipe at its junction with the first stage, and nailed to the stage. The underside of the hole in it should be rasped off to prevent the wood cutting the pipe when bossed over. This piece of oak is called a cleat. Cut out a piece of 8-lb. sheet lead of the same size as the cleat and drop it also over the pipe. Saw off the pipe about 1 in. above the lead and boss the end on to the latter. Fix another length of the delivery pipe and treat it in the same manner at the next stage, the bottom end being dropped into the tafted end on the first stage, thus making a block joint. The spigot end can be soiled and shaved above ground and let down to the plumber engaged on the job in the well. The socket end will need to be soiled and shaved in position, and the joint will have to be wiped down below. The metal pot should always be let down in a pail with great care, as serious injury would be caused to the man below if it overturned.

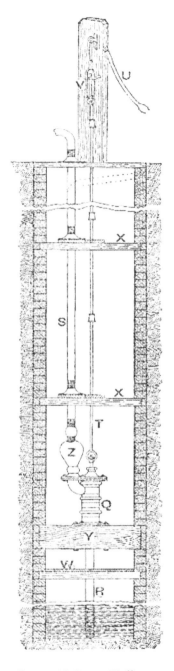

Figure 12 Deep Well Pump

In wet wells it is sometimes very difficult to prevent the water that soaks through the sides and drops from the stages from getting into the joint during the process of wiping. In such cases an umbrella can be used with advantage. Plumbers working in wells should wear a sack having a hole cut at the top for the head and a hole at each side for the arms.

The same process of fixing and wiping is repeated at each stage till the top is reached, the delivery pipe being either passed through the side and taken to a cistern, or passed through the top in the ordinary way. Sometimes a winch is used for raising the rods, and

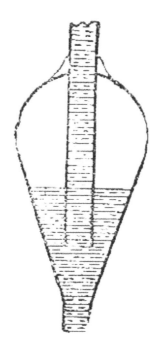

Figure 13 Air Vessel

sometimes a lever similar to that illustrated in Figure 12. The deeper the well the more the power that is needed at the handle. The ordinary lever will answer up to 60 ft., but when the distance becomes greater it is better to use a winch.

The rods are of iron, from 5/8 in. to 1 1/4in. diameter; 34in. and 1 in. are useful sizes. They are made in 12-ft. lengths. The bottom end of the rod that is connected with the pump is provided with a loop, as seen in Figure 12. The bucket rod, which is of copper, passes through a hole in this loop, and is fastened with a nut on top and a nut below, both screwing on the rod; this enables the rods to be regulated when desired.

The joint by which two rods are connected together is shown at Figure 14, the two ends being dovetailed and a brass clip hammered over them.

Fix the rods to the pump, keeping the bucket off the bottom valve, let the handle go right up, and cut the rod to suit the handle. It is essential that the handle should be kept right up and the bucket off the bottom valve, for if the bucket touched the valve, the valve would in all probability be bent or broken.

The rods should pass through the stages and through oak cleats in the same way as the delivery pipe; or they may be fixed on the face of the stages and passed through rollers. Both methods have their advantages and disadvantages, the rollers being liable to become clogged with rust and dirt. The

Figure 14 Pump Rod Joint

oak cleats have been found to last as long as the rollers, and to be much more effective.

The top valve can now be fixed, and the delivery pipe screwed on. This was not fixed before, in order that anything that might fall through the pipe during fixing might be prevented from entering the pump and getting under the valves. The pump can now be tried, and, if fixed according to the above directions, should work satisfactorily.

The repairs most needed for the above pump will be new leathers for the valves, which are the same as those shown at Figure 9, and are fixed as already explained.

To find the pressure to be overcome in raising water by means of a pump, multiply the square diameter of the barrel in inches by .34, which will give the number of pounds to be overcome for 1 ft. raised.

To find the number of gallons raised per hour, multiply the square diameter of the barrel in inches by the length of the stroke in inches, and then multiply by the number of strokes per minute multiplied by 60, and divide by 353. This will give the theoretical quantity raised. To determine the actual quantity raised deduct one-quarter from the theoretical quantity.

CHAPTER II

MAKING SIMPLE SUCTION PUMPS

A comparatively cheap, effective, and portable pump is represented by Figure 15; it can be made of No. 20 B. W. G. best galvanised iron, and the pattern for the barrel a is a rectangle, two sides of which are equal to the proposed length of the barrel, and two equal to the circumference plus 1 1/4in. for a seam lap. Set the compasses to 5/8 in. and mark along each length edge; and at intervals, along the lines thus marked, punch 1/8-in. holes, so that when the pattern is turned to shape the holes along one edge will correspond with those on the other. Remove the burrs and roll the pattern to shape through a pair of rollers, or, in the absence of rollers, bend it over a round stake.

The seam is next riveted together and sunk with the pane end of a hammer so as to render it smooth inside, otherwise torn bucket-leathers will result when the pump is used. The seam of the barrel should be soldered from the inside, care being taken to sweat the solder in the seam and over the rivet heads, rather than to leave a body projecting inside.

Raw spirits of salts (hydrochloric acid) is a good flux for soldering galvanised iron.

The method of setting out the pattern for the clack barrel B is shown in Figure 16. Here A a equals the depth of the pattern required, C D equals the larger diameter of the barrel, and E F the smaller. Draw lines from D to F, and from C to E, and produce them until they meet as at O. Then with O as centre and O A and O a as radii respectively, describe arcs of circles. Along the larger arc make G A and A H each equal to 3 1/7 times A D. Unite G O and H O by straight lines, the pattern required being G A H h a g.

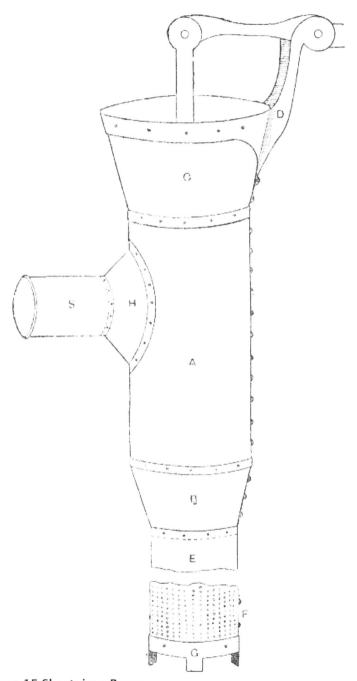

Figure 15 Sheet-iron Pump

The working edges, as shown by the dotted lines, are additional, and holes are punched for riveting as indicated, after which the pattern can be turned to shape and riveted together. The seam should be sunk and the working edge of the larger diameter should be set off slightly inward, while the edge of the smaller diameter is set off slightly outward, as shown in Figure 15.

It is much better to make all the parts before attempting to do any building up.

The method of obtaining a pattern for the spout

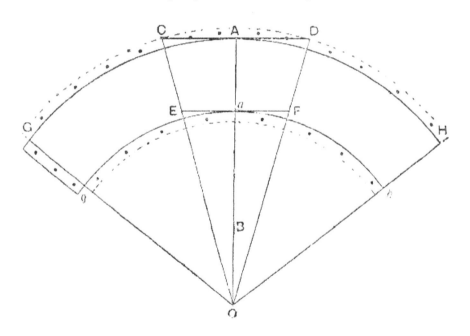

Figure 16 Pattern for Pump's Clack Container

S (Figure 15) is shown in Figures 17 and 18. Draw an elevation of part of the barrel and the spout as in Figure 17, and, to obtain the lines of interpenetration, draw a half-plan of the barrel as a, b', c', and with a as centre draw a quarter plan of the spout; divide this into three equal parts, as at 1, 2, and 3, and from these points draw lines parallel to a c', to touch the arc a b' c' in the points 1', 2', and 3', and from these points draw lines perpendicular to those already drawn. Next on a line A G at the end of the spout describe a semicircle and divide it into six equal parts as shown. From the points of division, draw lines at right angles to the line A G, cutting it in B, C, D, E, and F, and cutting the lines at the other end respectively in b, c, d, e, and f. A line drawn from a through these points to g gives the required junction.

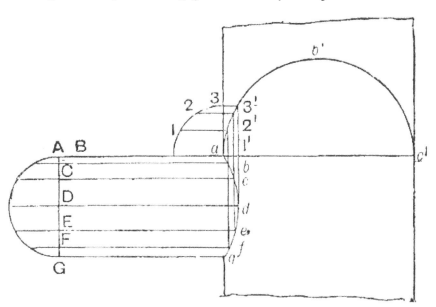

Figure 17 Setting out Spout Pattern

To develop the pattern for the spout, along G G' (Figure 18) set off twice as many equal parts as there are in the semicircle described on the line G A (Figure 17), and from these draw lines perpendicular to G G', and to them transfer from Figure 17 the respective lengths

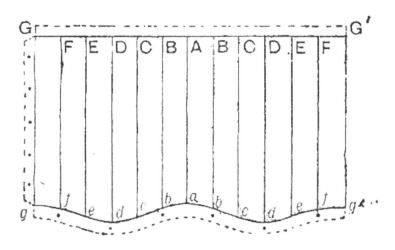

Figure 18 Setting out Spout Pattern

of the lines A a, B b, and so on. A connecting curve touching the ends of these lines gives G G' g g', the pattern required. Further additions for working edges must be made as shown by dotted lines.

Punch holes where indicated in Figure 18 for riveting, wire along the edge G G', turn the spout to shape, and rivet the seam. This should be soldered strongly from the inside, after which the edge which was previously allowed on the curved end of the spout should be set off outwardly so as to sit properly on the barrel. Place the spout in position opposite the seam of the barrel, and mark around the spout, and also the rivet holes. Punch the holes in the barrel, and an annular edge described inside this ring of holes encloses that part of a barrel which is now cut away.

A pattern for the top of the pump C (Figure 15), which fits over the barrel, can be obtained by adopting the principle explained in the case of Figure 16. It will greatly increase the strength of this top if the arm

D (Figure 15) has a hoop, so as to fit tightly around the top edge as shown, and the riveting edge of the smaller diameter of C should be set off outwardly to fit tightly over the barrel.

Figure 19 Pump Clack

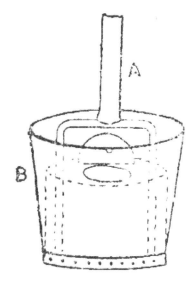

Figure 20 Pump Bucket

Patterns for the lower pipes E are not shown, as they are simply rectangles, two sides of which are equal to the proposed length of the pipe, one side equal to the proposed circumference plus a lap for the seam, while the other is 1/4 in. less, so that when rounded and riveted it will fit tightly inside another pipe. A hoop G with three short legs drives on the end of the perforated pipe F, and is riveted in position.

The clack (Figure 19) is a block of wood turned to fit

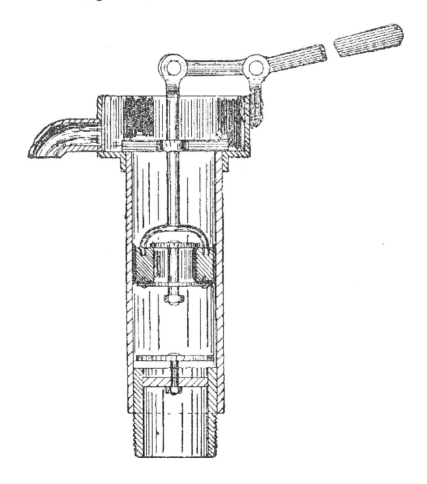

Figure 21 Yacht Pump

its chamber and having a large hole bored through it; this hole is covered at the top with a stout leather flap, weighted with a small block of lead, and working in a hinge-like manner, the leather being of sufficient size to form a hinge which is fixed with stout copper nails. The outside of the clack should be smeared over with a mixture of red- and white-lead and tow, and then driven into the galvanised-iron part. Four 1-in. copper nails driven through the iron and soldered over at the heads will securely hold it in its place.

The spout is next riveted and soldered to the barrel, and a boss H (Figure 15) can be riveted and soldered to both. The barrel is riveted and soldered to B, B to E, and so on to F and G. The iron arm, etc., D is riveted to C, which in its turn is riveted and soldered to the barrel.

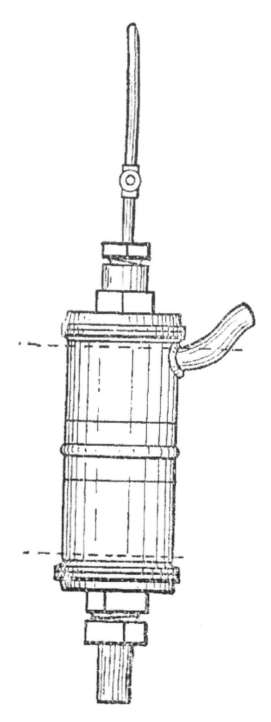

Figure 22 Beer Pump

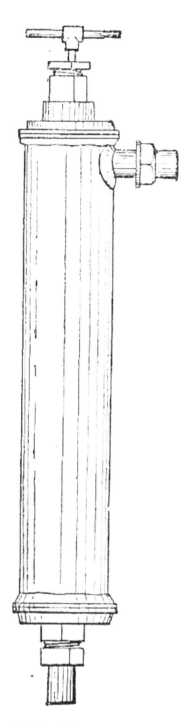

Figure 23 Yacht Pump

The wood bucket (Figure 20) has a central hole with a leather flap on top as previously described. The ironwork a pierces straight through the wood, and is secured underneath by two nuts and washers. The leather B is cut slightly taper, so that the top will be rather larger in diameter than the barrel, and is fastened to the bucket at the bottom with copper nails. It is better to have a lap of, say, 1 in. at the seam, which should be chamfered off so as to accommodate it to the sides of the barrel.

The handle J (Figure 15) can be attached to the pump and piston with short bolts and nuts, and the ironwork can be made by a local blacksmith.

Figure 21 shows a common suction pump suitable for a small yacht or other boat. To make it, obtain a piece of brass tube as thick and stiff as possible, whose internal diameter is 2 in.; this is to form the barrel of the pump. The top is made of sheet brass as shown, and soldered to the top of the barrel, which will be about 8 in. in length, the top to have a suitable spout attached that will reach over the side of the

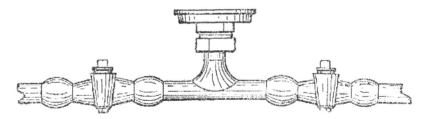

Figure 24 Two Taps on Suction Union

boat. The foot-valve seat at the bottom is made of a piece of stout brass tube, 2 in. external diameter, screwed half-way up, the other half of which is driven into the pump-barrel and soldered. About three-quarters of the distance from the bottom a cross-piece is soldered to form a guide for the stalk of the foot-valve; this stalk is 1/4 in. in diameter. A nut at the bottom of the stalk prevents the valve from having too much lift; 1/4--in. lift will be ample for this pump. If it is thought awkward to put the nut on from the bottom side, a bolt put in from the bottom with the nut on top perhaps may be more suitable.

The pump rod, which is forked at the bottom, ends in a double eye at the top. The lever is attached to this, working in a double eye as a fulcrum, which is riveted to the side of the top of the pump. The forked end should be driven tight into the bucket and soldered in its place, after the valve has been fitted to the bucket. The portion shown black all round and under the bucket is an ordinary pump leather which anyone can make for himself. This leather is held in its place by means of a brass disc fastened to the brass bucket by cup-headed screws. This brass disc has a hole in the centre, and forms a guide for the valve stalk that comes through the bucket and through a hole in the disc, being secured by means of a nut.

The valves themselves should be made of 1/4-in.

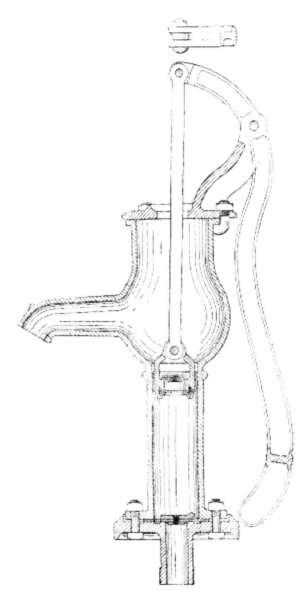

Figure 25 Small Cistern Pump

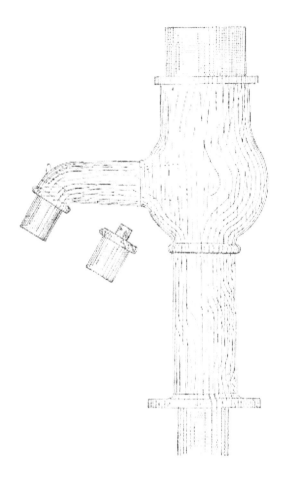

Figure 26 Elevation of Body Pattern of Cistern Pump

Figure 27 Print, Bead, and Fillet at End of Spout

brass plate, nice and level and faced up a little on the face-plate. The faces that they come against should be also something like true. The guide at the top of the pump is to be made in two pieces so as to clasp round the rod, after which it is to be sol-

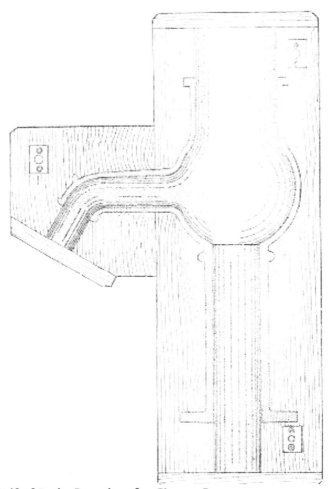

Figure 28 Half of Body Core-box for Cistern Pump

dered to the top. This pump is for shipping and unshipping at pleasure, for which purpose a socket should be made of brass and screwed to the deck between the gunwale and the edge of the cockpit, the socket to terminate in a pipe underneath the deck, to which a lead pipe 1 in. in diameter can be jointed and led away aft to the well. The bottom part of the pump, as will be seen, is for screwing into the socket; and when the pump is unshipped a suitable brass cap can be screwed into the socket to keep out dirt. The end of the suction pipe should be protected by a strum or grid to prevent chips and the like from coming up the pipe.

A pump for a small yacht may be made from an

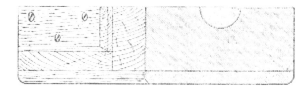

Figure 29 Section through Barrel Core of Body Core-box

old beer pump, such as is illustrated by Figure 22, which can be obtained for a few shillings. Select one of 21/2-in. bore, cut off the flanges as shown by the dotted lines in Figure 22, and also the tacks. Get a piece of 21/2-in. lead barrelling 1 ft. 3 in. long, trim it on a mandrel, square the ends, and solder the flanges on the top and the bottom. Then solder in a 1-in. union for the outlet, and the tacks on the back for fixing. Fit a longer rod to the bucket, and make a cross handle from parts of the old rod fittings, as in Figure 23. For convenient working, fit up the pump so that the handle comes about 2 ft. 6 in. above the floor. Such a pump can be fitted up for flushing a yacht closet. The spout is connected to the arm of the water-closet pan, and two taps are fitted to the suction union as shown in Figure 24. One is connected to the pipe leading to the outside water, and the other to the pipe leading to the bilge. Care must, of course, be taken that the taps are not both left open at once, or the boat will soon be filled.

A small cistern pump is shown in section by Figure 25, full particulars of the necessary patterns and core boxes used in casting it being given in the illustration

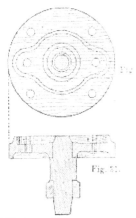

Figure 30 Pattern for Base of Cistern Pump

Figure 31 Pattern for Base of Cistern Pump

tions (Figure 26 to 41), the inscriptions to which clearly describe them. In making the body pattern, two pieces of wood are first dowelled together and temporarily fastened with screws for the purpose of turning in the lathe. A neck is turned at the lower end to receive the lug, which should be turned in halves. Two pieces may be fastened together and turned to form the spout, the bent part being worked by hand. The print, bead, and fillet, in halves, can be turned with a plug which is let into recesses made in the

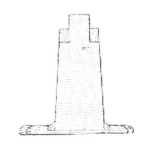

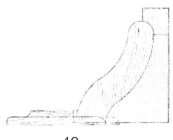

Figure 32 Side Elevation of Pattern for Cover

Figure 33 Front Elevation of Pattern for Cover

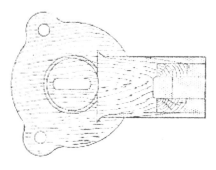

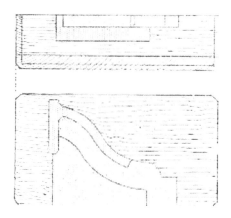

Figure 34 Plan of Pattern for Cover

Figure 35 Half Core-box for Coyer Pattern

Figure 36 Half Core-box for Coyer Pattern

pattern. The spout is attached to the body by a dovetail piece, a flat being cut on the body pattern and a piece glued on from which a radius is worked with the gouge.

In making the body core-box two pieces are

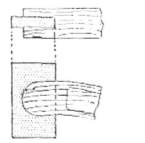 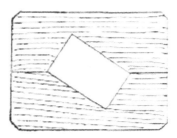

Figure 37 End of Pattern for Handle, showing Core Print

Figure 38 Core-box for Handle Pattern

dowelled together, a projecting part being temporarily attached to each piece for the spout core. Transverse backing is glued and screwed on to prevent warping, and one-half of the pattern is laid on the face of the box and its shape scribed around it. From this the thickness of metal is marked.

The section for the barrel core is then worked

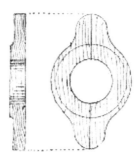

Figure 39 Pattern of Lugs and Fillet at Bottom of Barrel

through one-half of the box, an allowance of 3/16 in. to 1/4 in. in the diameter being left for boring. A template of the upper part of the section is made to work radially, and the upper part is cut out to it. For working the section of the spout core, the projecting part of the box may be removed after the shape is marked. One half of the box being worked, the other half may be marked out from it by means of a

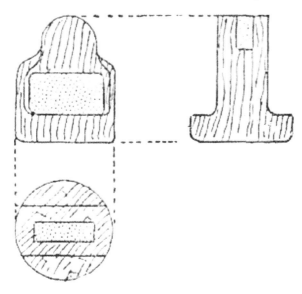

Figure 40 Pattern for Pump Bucket

stiff wire with the point bent. After being glass- papered the end pieces are screwed in place.

The cover pattern may be partially formed by turning. The slot for the pump rod and two of the holes for the bolts may be formed in the pattern to leave their own cores. If the bracket be moved

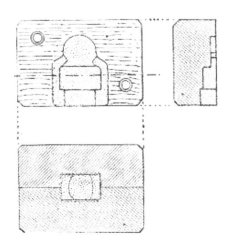

Figure 41 Half Core-box for Bucket

through 60° on the cover and the direction of the slot altered correspondingly, all three bolt holes may be so made. The print is arranged so that the mould may be jointed about half-way through the thickness of the plate and level with the top of the print.

The core-box to fill this print and core out the space at the back of the bracket is more easily marked out when made in halves, but as parts are undercut the core is removed from the box before the halves are separated.

The pattern for the bucket is turned solid and afterwards cut, the core from the corresponding box forming the interior.

The base pattern is moulded in a vertical position; the centre part is cored out, a deep bottom print and shallow top print being turned with the body of the pattern. Where the diameter is increased for receiving a screw thread, a loose wood collar is left which necessitates an extra joint in the mould. This may be avoided by cutting the collar into several pieces which draw into the mould, or by putting a print around the diminished section of the circular stem and forming this part by means of a core.

The pattern for the handle is made exactly like the casting, except that at the upper end a small print is formed for the jaw.

CHAPTER III

PUMP CUP LEATHERS

As the efficiency and satisfactory running of a pump largely depends on the proper packing of the bucket, and as cup leathers are almost exclusively used for this purpose, too much stress cannot be laid on the necessity for having leather of the best quality only, and using it so that the fibres may not be unduly strained in pressing.

For making cup leathers, metal moulds and stamps, or pistons, will be required. If only a few cup leathers are to be made by a rough method, very hard wood moulds may answer instead of metal. The cups are made of the best oil-dressed leather, 3/16 in. or 1/4 in. thick, steeped in warm water at about 180° F., and each is forced gradually into the mould by means of a bolt and nut, and left till it is again hard. The leather should be cut out of the back in preference to the belly or limbs of the hide. If desired, the leather can be planed down to 1/8 in. The flesh side should be outwards. The leather should be cut circular, the diameter being the diameter of the pump plus 11/2 in. to 3 in., according to the size of the pump, to form the cup sides. The leather is placed on the mould as shown, and then slowly forced down with a lever, screw, or other mechanical appliance. A different mould must be used for each size of pump. The centre of the bottom is cut out after the leather has been pressed and the top edges of the cup are trimmed.

Figures 42 and 43 are section and plan of a moulding press, a is a cast-iron mould, with a recess bored to the outside diameter of the finished cup, and on the outside are the two lugs b to screw the press to a bench or table, two 1/8 -in. holes in the bottom of the mould allowing the air to escape while the leather is forced down. The centre of the mould takes the 5/8-in. bolt C; D is an iron block turned to the inside diameter of the finished cup. A piece of leather is shown forced into shape.

Dress the cups after moulding with the following: —

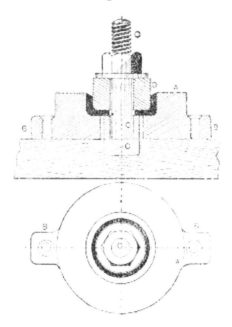

Figure 42 Section of Pump Cup Leather Moulding Press

Figure 43 Plan of Pump Cup Leather Moulding Press

Four parts of best linseed oil, 2 parts of olive oil, 1 part spirit of turpentine, 2 parts of castor oil, 1/2 part of beeswax, and 1/4 part of pitch. Boil these together over a gentle fire in an earthenware vessel, and during ebullition dip in the leather, and let it remain in for a few minutes, fifteen minutes being sufficient for thick leathers.

If the leathers are required to be hard—almost like vulcanite when finished, use the ordinary best sole leather, and soak it for three days in a bath composed of slaked lime 3 lb., to 11/2- gal. of water. Then pass it through water heated to about 180° to 190° F., and while hot press it in the mould. The finished cups may then be dressed with the hot lubricant.

To understand the requirements of a suitable leather for pumps, an investigation of the action going on in a pump barrel is necessary. Figure 44 is a section

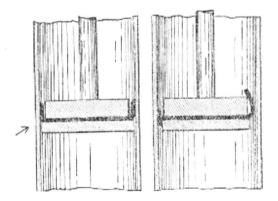

Figure 44 Section of Pump Cup Leathers and Buckets

Figure 45 Section of Pump Cup Leathers and Buckets

of a pump leather and bucket, the cup leather being indicated by the thick black line. Assuming that the bucket is making its up-stroke, the water above the bucket will press the ends of the leather against the inside of the barrel, making a perfectly water-tight joint. There will be a certain amount of friction between the leather and the barrel. The greatest friction will be at the point indicated by the arrow, the amount of friction depending on the pressure and the quality of the leather. Cup leathers at this point should be well supported, as shown by a clamping ring, which should be a good fit in the barrel, with just sufficient clearance to allow the bucket to work up and down freely without side play. Some cheap, badly designed pumps have this clamping ring much too small, there being in some cases as much as 1/4-in. clearance between the bucket and the wall of the barrel; many have even more than this. With such pumps, the leather, not being supported at the point of greatest pressure, after a little time is forced between the bucket and the barrel, which ultimately jams the bucket on the up-stroke, often to such an extent that it is impossible to move the bucket either up or down without exercising great force. This fault is shown on the left-hand side of Figure 45.

Another important point is the depth of the leather. With a leather too deep, the top edge bends inwards, as shown on the right of Figure 45, and the water gets between it and the barrel wall, forcing the leather away instead of making a perfectly water-tight joint.

The following table of thickness of leather and depth of cup for different diameters is based on experience, and if followed will give an ideal cup for lifts up to about 100 ft.

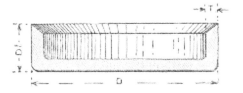

Figure 46 Diagram showing Proportions of Pump Cup Leather

Diameter.	Thickness.	Depth.
2 in.	3/16 in	5/8 in.
3 in.	3/16 in.	3/4 in.
4 in.	1/4 in.	1 in.
5 in.	1/4 in.	1 1/8 in.
6 in	1/4 in.	1 1/4 in.

Cups of these dimensions make a perfectly water-tight joint, and the friction is reduced to a minimum. Figure 46 shows the proportions of thickness T, depth D, and diameter D of a 3-in. leather.

There is no advantage in having a deep cup, and undoubtedly a shallow leather is preferable. It works better than a deep one, and in moulding considerable less strain is required. Consequently the fibre at the sides is much stronger. With a deep cup the fibre is practically destroyed in pressing, on account of its having to be pressed to a greater depth than is really necessary.

It should be borne in mind, when pressing, that the pressure should be evenly distributed over the leather. This is impossible when employing a rough- and-ready method of pressing. To obtain an even pressure all over the surface of the leather, the pressure will have to be applied at the centre, as shown in Figures 42 and 43, these indicating the method generally employed in the manufacture of cup leathers.

Some people prefer cup leathers with the skin side outwards, and others with the flesh side outwards. Cup leathers with the flesh side outwards last longer, and make a more perfect water-tight packing; this no doubt is due to the outer stronger fibres of the leather not being worn away, as is the case when the skin side is outwards.

When putting in a leather, it is very important that it should be well lubricated with tallow, which should be well worked into the leather with the hands. Another way of lubricating is to soak the cup in neats- foot, sperm, olive, or castor oil, for some time before putting in. In no case use mineral oil, as this will make the leather rotten and pulpy.

Ordinary machine oil is largely composed of mineral oil, and therefore should not be used to lubricate the leathers.

CHAPTER IV

PUMP VALVES

In pump-work the valves are the weak points. The duty of a valve being to allow the passage of a fluid or liquid in one direction only, the parts in contact when the valve is shut should fit so exactly as to prevent any of the liquid passing through them. If one of the faces is made of some elastic pliable material, such as india rubber, it will adapt its form to that of the hard face with which it is brought in contact, so forming an air- or water-tight valve; but if both the valve and its seat are hard and unyielding, the surfaces of contact must be made exact counterparts of each other in order that they may work satisfactorily.

All surfaces of special forms may be fitted by grinding together with fine emery and finishing off with crocus powder, but to flat surfaces a different process should be applied.

The mechanical method of preparing a plane surface is as follows: —In the first place, the surface is made as true as the planing machine or lathe will render it. Then the surface is tested by laying it upon a rigid plate having a true surface, on which a little ruddle has been smeared; the plate used for this purpose is called a surface plate, and is one of the most important gauging tools in the shop. Any parts in the surface under manipulation that are higher than the general level will be marked with the ruddle if the work be moved slightly. The work being then turned over on to the bench, the marked parts are carefully scraped down (a very good scraper can be made from an old three-square file, by grinding the cuts out at the end so as to form three curved scraping edges—not cutting edges) and the work is again applied to the surface plate. More parts will now be found to be marked, and these are to be scraped down as before; at every application it will be found that the marking will extend more continuously over the surface under treatment until at last a uniformly even

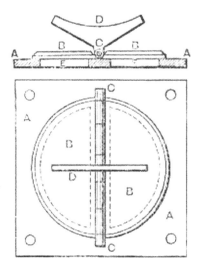

Figure 47 Double Flap or Butterfly Valve

Figure 48 Double Flap or Butterfly Valve

surface is produced, and both the valve and its seat being thus prepared, a perfectly tight contact may be anticipated.

For very large surfaces it becomes necessary to apply the surface plate to the work, small ones, easily handled, being made for this purpose. As perfect rigidity is absolutely indispensable in surface plates,

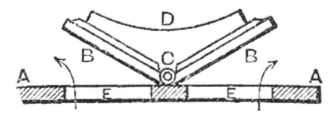

Figure 49 Butterfly Valve, Open

they are strengthened by being made with deep ribs at the back, running both longitudinally and across, which impart the necessary quality.

One of the oldest forms of pump valve is the double-flap or butterfly valve shown in Figures 47, 48, and 49. The valve seat is shown in section in Figures 47 and 49 to exhibit the apertures through which the liquid passes when the valves are open (see Figure 49), the liquid passing in the direction of the arrows; A is the valve seat, having in it openings, E, shown dotted in the plan, Figure 48; B valves which work upon hinges C, their movement being limited by the guard D; the letters apply to the same parts in each figure; the valves are semicircular in plan. Whenever the pressure below these valves exceeds that above them by the amount of the weight of the valves, they will open, closing again as the pressure becomes equalised on each side. For small pumps with a light load of water, these valves serve well enough, and are, therefore, suitable for lift pumps worked by hand at slow speed; but, at a high velocity, the rapidity and violence of their closing would soon damage the sea tings and render them leaky if made with metal seats and valves; the valves may, however, be lined with leather, provided the pressure of water is not great. In lift pumps, the pressure will not exceed 15 lb. per square inch in addition to that of the water above the valve in the pump bucket.

The weight of a cubic foot of water is almost exactly 62 1/2lb., therefore the height of a column of

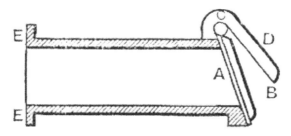

Figure 50 Flap Valve, Shut

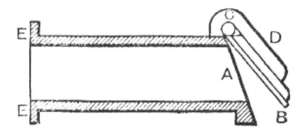

Figure 51 Flap Valve, Open

water pressing upon a valve being known, the pressure per square foot upon it will be that height multiplied. by 62 1/2 lb., and the pressure per square inch will be equal to the height in feet multiplied by 125 and divided by 288.

The area of escape for the liquid around the edges of the valves will be equal to that of the openings E, when the lift of the outer edge of the valve equals two-fifths of the diameter of the openings E.

As stated above, the valve will not open until the pressure of the fluid below exceeds that above it by an amount equal to the weight of the valve. In pumps for water and other liquids, this only requires consideration in the case of suction pumps, in which the weight of the column of water, in addition to that of the valve, must not exceed the pressure of the atmosphere; but for exhausting pumps—such, for instance, as the air pump of a steam engine, or the pumps for exhausting the sugar refiners' vacuum pans—more delicate arrangements must be adopted.

If a vacuum is to be maintained, the valve must

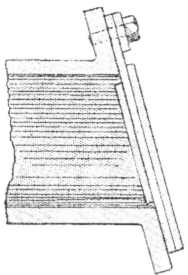

Figure 52 Leather-flap Valve

be of the least possible resistance consistent with its closing tightly enough to prevent a reflux of air, gas, or vapour, as the case may be. In Figures 50 and 51 is shown a flap valve, the seat and suction passage

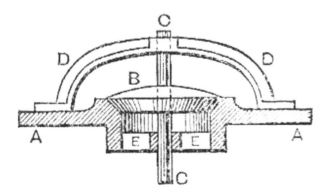

Figure 53 Stem Valve. Shut

being in section, in which the resistance to opening is determined by the angle of the valve seat to a vertical line. If the valve b, depending from the hinge c, were hung vertically, it would exercise no pressure upon the valve face a, and as the inclination of a increases, so will the resistance of the valve b; d is a guard to restrict the opening of the valve as shown in Figure 51; e, e are flanges for attachment to the vessel to be exhausted.

Details of a leather flap valve are given in the sectional view (Figure 52). Such a valve consists of a strip of leather stiffened with metal plates.

A spindle valve, having a conical seat, is shown

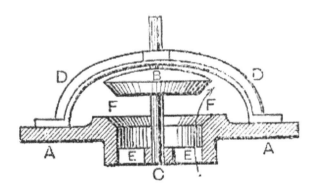

Figure 54 Stem Valve, Open

shut in Figure 53 and open in Figure 54. This valve requires to be ground into its seating to make a true fit. The seat a is shown in section, f being the part upon which the valve closes; B, the valve, carries a rod or spindle C, of which the lower end works through a hole in the centre of a wheel-shaped frame at the bottom of the inlet, between the spokes of which are the openings E to allow of the passage of liquid; the

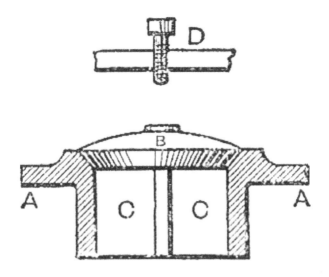

Figure 55 Stalk Valve, Shut

upper end of the spindle passes through a hole in the guide d, which also serves as a stop to restrict the rise of the valve B; the height of rise necessary to afford an outlet equal to the inlet pipe is one quarter the diameter of the latter. These valves are simpler than the flap valve with guard, inasmuch as the hinge is dispensed with; they also do not require to open so widely in proportion to width as the flap valve,

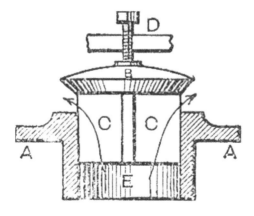

Figure 56 Stalk Valve, Open

and will therefore close quicker and with less concussion. Rapid closing in valves has some special advantages which will subsequently be referred to in connection with a larger class of valve.

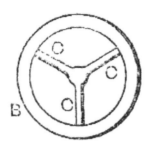

Figure 57 Underside of Stalk Valve

A very compact form of valve—stalk valve— similar in principle, but more self-contained than the above, is illustrated in Figures 55, 56, and 57, the seats in the two former figures being shown in section. Figure 57 is a plan view of the under side of the valve. This valve, which, on account of its compactness, has been much used for feed pumps, and others that are re-

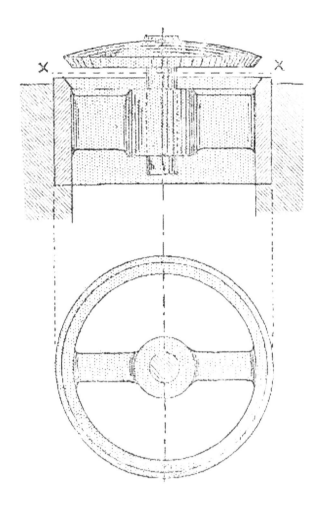

Figure 58 Conical-disc Valve

Figure 59 Conical-disc Valve

quired to be stowed away in small spaces, is made with a stalk consisting of three wings or ribs cast on its under side; a is the valve seat, which has its conical part ground to fit the edge of the valve b, and is truly bored out to receive the wings C, the edges of which are turned to fit it, and thus act as a guide to the valve. The rise of the valve in this case may be restricted by a set-screw D passing through the top of the valve box. The liquid escapes, when the valve is open, between the wings C as shown by the arrows in Figure 56.

A valve similar in principle to the two last described is shown by Figures 58 and 59. This is the conical disc valve, the lift of which should not exceed one-fourth the diameter of the valve. The central spindle acts as a guide, and the lift is limited by a stop.

A simple kind of valve consisting of a ball B in a spherical seating ground in A at the top of the inlet E is shown in Figure 60; the ball is kept in place by a guard D. Where considerable area is required it has been proposed to agglomerate a number of these

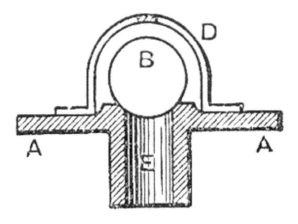

Figure 60 Ball Valve

valves. Ball valves must necessarily be disproportionately heavy in comparison with their areas; the balls also are somewhat costly, as they must be perfectly true to be of any use at all, and taken altogether they do not seem to possess any particular advantages to recommend them. Technical details of a ball valve are shown by Figures 61 and 62, the former figure representing half vertical section and half elevation. The lift of the ball is limited to 1/4in.

Passing from the comparatively small purposes for which the valves above described are appropriate to others of greater magnitude, some essentially different forms of construction are met with. In order to make clear what is required of the valves under different circumstances, it is necessary to advert to the actions of the different kinds of pumps to which they are fitted. The simplest form is the common suction pump, with a flap or butterfly valve at the top of the suction pipe, and another in the pump bucket itself. It occurs most frequently in agricultural districts where there is no laid-on supply of water, though the principle has also formerly been used in connection with large draining schemes. Next is the lift or force pump with a valve at the outlet to prevent the water discharged under pressure from returning; but this is displaced by the plunger force pump (described in the next chapter) in which pump, instead of a packed bucket or piston to draw and force the water, a long plunger is used working through a water-tight stuffing-box in the top of the pump barrel. Double-acting pumps have pistons worked by rods passing through stuffing-boxes in the pump covers, so that work is done simultaneously on both sides of the piston: drawing on one side and forcing on the other. For lifting and forcing large volumes of water the single-acting Cornish engine has been very extensively used, being one of the most economical in its action, and therefore generally adopted before the era of very high pressures commenced. In these machines a plunger pump is used, having on its rod a box loaded with weights, which in its descent forces the water out of the pump barrel; the pump plunger, together with its " preponderating weight, " is lifted by steam pressure on one side of the piston, the other being open to the condenser. The exhaust is then shut and communication made between the spaces above and below the piston, thus equalising the pressure and leaving the pump load free to descend. The pump barrel is filled through the suction valve during the steam stroke. These pumps naturally work very smoothly; as there is no sudden change of motion, the plunger comes steadily to rest, allowing the barrel to fill completely before the down or forcing stroke is made, and if the pump valves act perfectly a full barrel of water is discharged at each stroke.

The valves for such pumps will necessarily be very large to allow the free passage of the water, and the old rule of making the suction pipe one half the diameter of the pump barrel soon became obsolete when the construction of large pumps came into the hands

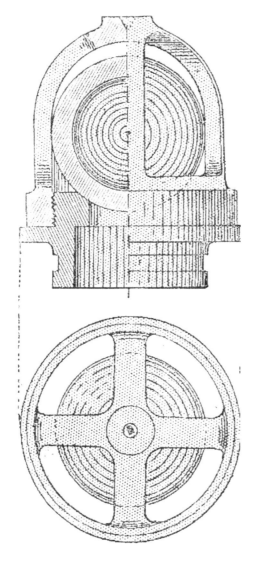

Figure 61 Ball Valve

Figure 62 Ball Valve

of thoughtful engineers, among whom Thomas Wick- steed stood pre-eminent as an improver of the Cornish engine. To give an idea of the mass of water dealt with, take a pump with a plunger 50 in. in diameter and working with a stroke of 11 ft. The reader will know that the area of a circle is equal to the radius multiplied by half the circumference, and the circumference is the diameter multiplied by 31/7; therefore the area of the plunger section will be 25 multiplied by 50 and by 31/7, and divided by 2 equals 1,964 sq. in., or, dividing by 144, equals 13 5/8 sq. ft.; multiplying this by the stroke of the plunger, 11 ft., we have 150 cub. ft. of water expelled at each down stroke of the plunger; as 1 cub. ft. contains 6 1/4 gal., this amounts to 937 1/2 gal. per stroke.

The friction of water in pipes varies as the square of its velocity, and the velocity of a given quantity of water passing through a pipe in a given time varies inversely as the square of the diameter of the pipe, therefore the friction of the water will vary inversely as the diameter of the pipe multiplied into itself three times—thus, if one pipe is half the diameter of another, the friction in the former will be sixteen times as great as in the latter; this shows the importance of making the valves and passages as large as possible.

These large pumps were at first fitted up with butterfly valves, but their unsuitableness to such a purpose soon manifested itself. The great area of the valve met such resistance in shutting that instead of falling through the water it hung until the return stroke, and fell with the column of water. In such a case as that taken above, where the pump acted against a pressure of 100 ft. of water, the weight thus falling would amount to 43 1/2 lb. on every square inch

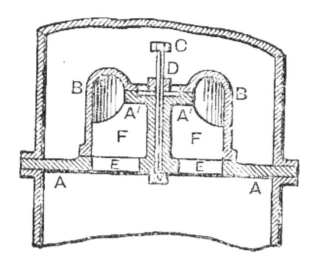

Figure 63 Vertical Section of Double Beat Valve. Shut

of valve surface, and would cause a concussion at every stroke that would speedily destroy the valve. The first method tried to obviate this difficulty consisted in providing air-cocks to admit air under the valves so as to allow of their falling more rapidly by their own weight; but this was obviously introducing a gratuitous leak into the apparatus, and in some cases as much as 16 per cent. of water was lost at

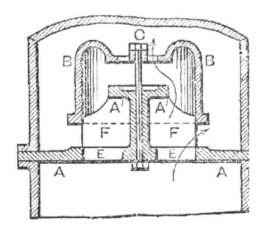

Figure 64 Vertical Section of Double Beat Valve. Open.

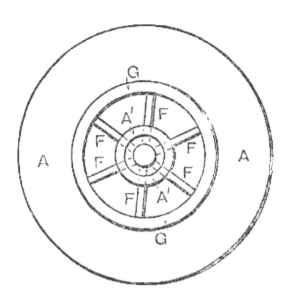

Figure 65 Plan of Seats of Double Beat Valve

each stroke, causing, of course, a corresponding waste in consumption of fuel. The real remedy was to be found in designing a valve that should close quickly from its own form without requiring any extraneous aid.

To get at the principle involved, consider the difference of the retarding force of a plate falling flat and one falling edgewise in water. The force tending to buoy it up and retard its passage through the water arises from the friction of the water passing from the under to the upper side of the plate, and when that plate is flat it evidently takes the water a longer time to reach its margin than when it is placed on edge; there is, in short, a much greater lateral movement of the water necessary in the former than in the latter case. Suppose the plate to be 1 in. thick and 1 ft. square, laid flat, when it has displaced its own bulk of water once it will have fallen 1 in.; if on edge when the plate has displaced its own bulk of water once it will have fallen 12 in.

The conclusion arrived at from the above is that, while keeping a sufficient area of opening of the valve

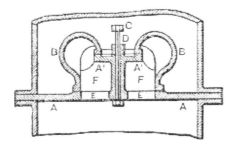

Figure 66 Vertical Section of Crown Valve, Shut

to allow of the free passage of the water, the area of the valve acting vertically must be reduced, and this is done by using the double beat valve shown in Figures 63, 64, and 65. Figure 63 is a vertical section of the valve closed, and Figure 64 of the same when open. A A' is a casting bearing two seats, of which the smaller A' is at the top; these are shown in plan at Figure 65, where G shows the lower and A' the upper seat. These are connected by ribs F, etc.; through openings E between these ribs the water passes from the underside of the valve B in the direction shown by the arrows. Part of the water passes under the lower edge of the valve B and the remainder over the top seat and out through openings round the centre of the crown of the valve. C is a nut which limits the rise of the valve, the hub D coming in contact with it when the maximum rise allowed is reached. This valve affords a very large outlet for the passage of the water, for there is the full area of the bottom seat, and the resistance to the escape is also relieved by the outlet at the top; the pressure being thus rapidly reduced within the valve also aids in facilitating its fall directly the flow ceases. The difference of area of the upper and lower seats will be that opposed to the water, and on this the whole weight of the valve is taken; so this form will require a greater pressure to open than a flat valve, but that is of no moment in comparison with the advantages obtained by its use.

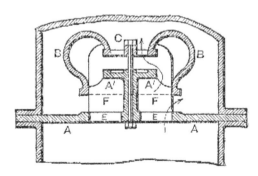

Figure 67 Vertical Section of Grown Valve, Open

These valves are made up to weights of 13 cwt., and sometimes more.

Under very heavy lifts the beat of these valves may be felt at some distance, and in some pumps which were used to force water to a height of 270 ft., crown valves of the form shown in Figures 66 and 67 have been employed; the letters refer to the same parts as in Figures 63 and 64, the only difference being in the form given to the valve itself; the idea was to give the valve an elastic form, and also to provide better passages for the water to the upper seat, A'.

As soon as it appeared that something different from the old-fashioned valves was needed, numerous patented inventions appeared. One of these consisted of a series of rings placed one upon another, and having a small disc valve on the top of all, the rings gradually diminishing in diameter upwards. The bottom ring formed a seat for the second, which formed a seat for the third, and so on to the top.

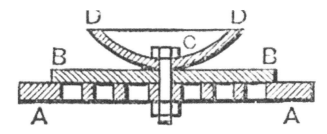

Figure 68 Flat India- rubber Valve, Shut

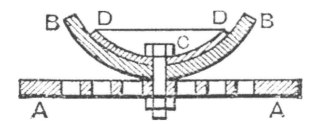

Figure 69 Flat India- rubber Valve, Open

This valve closed quicker than a plain disc valve of equal diameter would do, because the water getting away between the rings did not take so long to get clear of them as to get from under a flat plate.

With the increasing use of high pressure, steam, and high speed engines, high speed pumps have also been introduced, the advantage being found in the employment of smaller and less expensive machinery to do a given quantity of work; therefore valves suitable to quick-working pumps have become a necessity. In such pumps there are necessarily all kinds of concussions, for their action may be said to comprise a

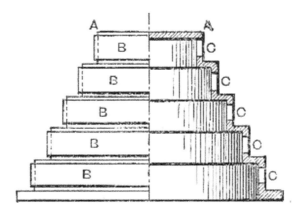

Figure 70 Indiarubber Band Valve. Half Elevation and Half Vertical Section

constant succession of sudden and violent reversals of motion; and if the engine is overrun actual blows will result, for there is a limit to the velocity for the suction end of the pump, which is that of the entering water. If this is exceeded the pump barrel does not fill, and on the return stroke the entering water is struck by the piston or plunger. In order to get the full advantage of the atmospheric pressure, it is desirable to place the pump below the level of the supply, for every foot of lift in the suction diminishes the speed at which the pump may be worked; whilst, on the other hand, the water can be forced out as quickly as the power at disposal will allow. With such machines it is evident that the natural fall of the valve will be too slow to meet the requirements of the case; the valves, therefore, must be shut by the returning current, the very action which

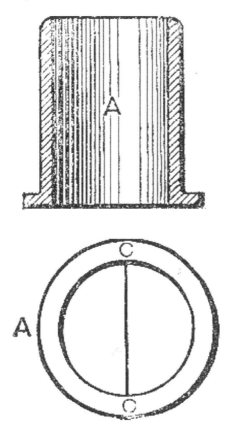

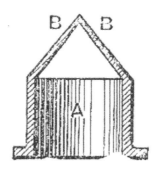

Figure 71 Indiarubber Semi-lunar Valve

Figure 72 Indiarubber Semi-lunar Valve

Figure 73 Indiarubber Semi-lunar Valve

in the large pumps proved fatal to the use of flap valves. In this case then the valves must work in the way required without the resulting concussions becoming destructive. The essentials are light and elastic material for the valves, large area of inlet and outlet for the water, and no working parts to shake or wear loose. The material that most nearly fulfils these conditions is indiarubber, but it has no strength to support the pressure of liquid over a considerable surface, therefore when used it is necessary to make the valve seat in the form of a grating, so that the indiarubber itself may be sustained against the pressure upon it. The less the lift of these valves the better, as the force of closing is proportionally less; therefore a large area, afforded by a considerable number of small valves is desirable.

A circular indiarubber valve upon a grated seating is shown in vertical section shut by Fig. 68 and

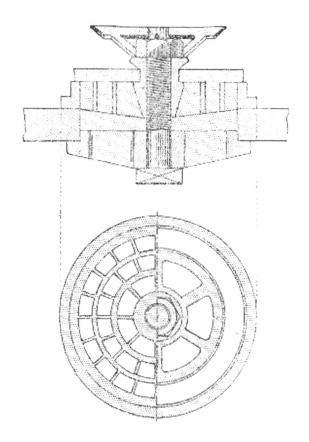

Figure 74 Indiarubber Disc Valve

Figure 75 Indiarubber Disc Valve

open by Figure 69. a is the seat and b the valve, which is bolted between the seating and a cup- guard D and secured by a nut C; the guarding D limits the rise of the valve and prevents its being torn or strained by any unusual pressure under it.

A form of valve in which, indiarubber bands are used was designed many years since for use with a Cornish engine raising water to a height of 100 ft. in one of the London waterworks. This valve, which gave perfect satisfaction, is shown at Figure 70, the left- hand half being shown in elevation, the right in vertical section. This arrangement consists of a series of diminishing cylindrical seats A A' in the form of vertical gratings having apertures C, around which are wrapped indiarubber bands B; the water pressure within the valve distends these bands and so escapes around their edges, and the bands immediately contract upon their seats on the relaxation of the interior pressure.

Muddy and gritty water will invariably derange metal-faced valves, inasmuch as some deposit is sure to get upon the seating; to obviate the inconvenience thus caused, an indiarubber valve of the form shown in Figures 71, 72, and 73 has frequently been used with success; it is, however, only applicable upon a small scale. It consists of a tube a, which is shown at Figure 73, in vertical section through the line C C (Figure 72), on plan at which the lips meet. Figure 71 shows a vertical section at right angles to the line c c at which the lips b meet; these lips are forced apart by liquid passing through, and it is obvious that mud and other clogging material will pass freely through, there being no valve seats on which to deposit. These valves are not suitable for high lifts on account of the yielding nature of the material of which they are composed. The idea appears to be taken from the form of certain valves occurring in the circulatory system of animals.

An actual representation of an indiarubber disc valve is given by Figures 74 and 75, the former figure being a vertical section and the latter—to the right— a half-plan of the complete valve and—to the left—a half-plan with rubber and grating removed. The rubber is bolted to the grating at the centre.

CHAPTER V

RAM OR PLUNGER PUMPS

A diagram showing the principle of a ram or plunger pump is presented by Figure 76, which is uniform with the diagrams of suction and lift pumps. (Figures 1, 2, and 11, pp. 10, 11, and 19.) In Figure 76 A indicates the suction pipe, B the pump barrel, H the ram or plunger, D the admission valve, F the delivery pipe, and G the outlet valve. The action of the pump shown in the diagram is exactly the same as that of the single-acting lift pump, except that there is a piston in one and a ram or plunger in the other.

The practical cause of failure of the solid-piston single-acting lift pump was its liability to jamb in the working barrel by sand and gravel, and leakage of its leathers was apt to cause serious accidents to the costly pump-rods, and even to the steam engine employed to drive the pump.

Plunger-poles fitted to the case—the latter having longitudinal grooves for the passage of the water— were used at an early date in France. One of these pumps described in a book published in the year 1827 illustrated by Figure 77, and consists of a wooden trunk, square or round, open at both ends, and having a valve at the bottom. O is the surface of the water in the pit or well, and K the place of delivery. The pit must be as deep in water as from K to O. B is a wooden trunk, open at both ends, and having a valve, P, at the bottom. The top of this trunk must be on a level with K, and has a small cistern E F. It also communicates laterally with a rising pipe G, furnished with a valve H opening upwards. M is a beam of timber, so fitted to the trunk as to fit it without sticking, and is of at least equal length. It is suspended by a chain from a working beam, and loaded on the top with weights exceeding that of the column of displaced water.

Sir Samuel Morland, Master of Mechanics to King Charles II., in the year 1674 invented and patented the plunge-pump made of cast-iron. One of the most

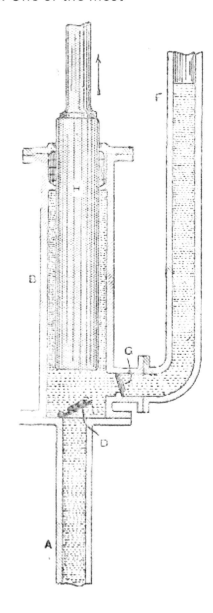

Figure 76 Plunger Pump: Explanatory Diagram

important features in plunger-pumps and dry-spear- pumps, as also in the steam engine, is the gland and stuffing-box, which was undoubtedly invented by Sir Samuel Morland, although it has been erroneously attributed to James Watt by some writers. The arrangement used before the invention of the gland and stuffing-box was described to the following effect in the " English Encyclopaedia, " published in 1802: —In forcing-pumps, instead of a piston, which applies itself to the inside of the barrel, and slides up and down in it, there is a long cylinder, nicely turned and polished on the outside, and of a diameter less than the inside of the barrel. The cylinder is called a working barrel, and is constructed as follows. The top of the barrel ends in a flange, pierced with four holes for receiving screw-bolts. There are two rings of metal of the same diameter, and having holes corresponding to those in the flange; four rings of soft leather, of the same size and similarly pierced with holes, are well soaked in a mixture of oil, tallow, and a little resin. Two of these leather rings are laid on the pump flange, and one of the metal rings above them. The plunger is then thrust down through them, by which it turns their inner edges downwards. The other two rings are then slipped on at the top of the plunger, and the second metal ring is put over them, and then the whole is slid down to the metal ring. By this means the inner edges of the last leather rings are turned upwards. The three metal rings are now forced together by the screwed bolts, and thus the leather rings are strongly compressed between them and made to grasp the plunger so closely that no pressure can force the water between them. The upper ring just allows the plunger to pass through it, but without any play, so that the turned-up edges of the leather rings do not come up between the plunger and the upper metal ring, but are lodged in a little conical taper, which is given to the inner edge of the upper plate, its hole being wider below than above. It is on this trifling circumstance that the extreme tightness of the collar depends. To prevent the leathers from shrinking by drought, there is usually a little cistern formed round the head

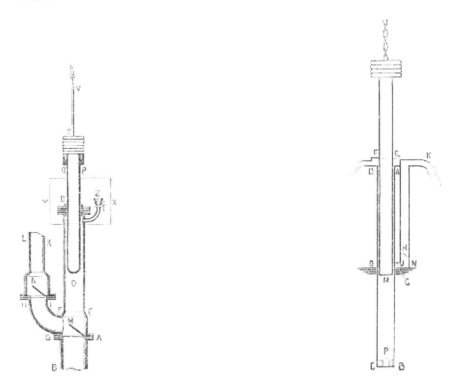

Figure 77 Ancient Ram Pump   Figure 78 Ram Pump with Delivery Clack on Side of Working Barrel at Bottom

of the pumps, and kept full of water. The plunger is worked either by a rod from a working beam or by a set of metal weights laid on it.

The above pump is illustrated by Figure 78, in which B is the top part of the suction-pipe, D the plunger-case, P the plunger or plunger-pole, V the plunger- rod, T the weights, M the suction valve, N the delivery valve, H the delivery clack-piece, L the delivery-pipe or rising main, Z is a cock for discharging the air which accumulates in the top part of the plunger-case, and Y is a cistern filled with water to prevent the packing leathers from getting dry.

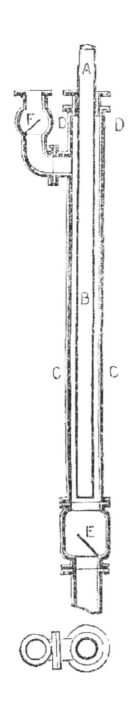

Figure 79 Sections of Trevithick Ram Pump

Figure 80 Sections of Trevithick Ram Pump

Plunger pumps include the following classes: — a. Single-acting plunger pumps, b. Single-acting hollow ram or plunger pumps, c. Double-acting plunger pumps, d. Bucket and plunger pumps. e. Piston and plunger pumps. f. Plunger and plunger pumps.

a. Single-acting plunger pumps may have the suction valve in a line with the plunger, and the delivery valve on the side at the bottom of the plunger case. A pump of this type is illustrated by Figure 78, and has already been described.

Again, they may have the suction valve in a line with the plunger, and the delivery valve on the side at the top of the plunger-case. This class of plunger- pump was designed by the great Richard Trevithick in the year 1797. It is illustrated in Figures 79 and 80. The plunger-pole was made of cast-iron, turned on the outside, and working in a cast-iron plunger-case or working barrel, the sides of which were not touched by the pole. The pole case was provided with a stuffing box and gland and the necessary valves. Trevithick's pumping gear was quickly appreciated, and during the succeeding four or five years many of the principal mines in Cornwall had their old bucket- lifts removed to make room for the new plunger-sets. This form of pump is still largely used—almost as Trevithick designed it in 1797. Figure 79 is a longitudinal section, and Figure 80 a sectional plan of the pump. A is the wooden plunger-rod, fastened into the hollow cast-iron pole or plunger B; C, the pole-case, allowing space for the passage of water round the pole; D is the stuffing box; E the bottom or suction valve, allowing the water to ascend into the pole case on the ascent of the plunger; F, the top or delivery-valve, through which the water is forced upward, through the pump-trees or rising main, on the descent of the plunger.

A third kind of single-acting plunger-pump has the clack pieces on the side of the plunger-case, so that the wind bore, suction clack, delivery valve, and rising main are all in one line. A sectional elevation of this pump is given by Figure 81. This arrangement is mostly used at the present day for convenience of fixing in the shaft or well, and because of the easy access to the valves. It consists of the plunger-case A;

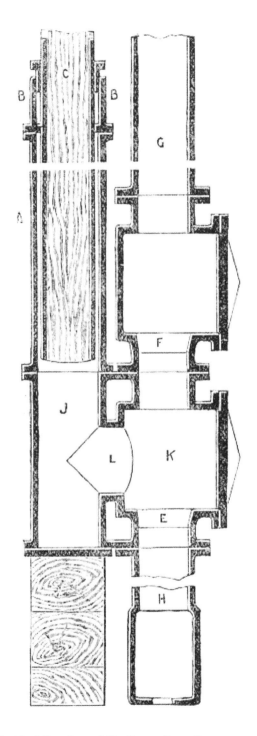

Figure 81 Vertical Section of Ordinary Ram Pump

the stuffing box and gland B, bolted to the top of the plunger-case; C is the plunger-pole, consisting of a cast-iron tube, turned on the outside, and forced on to a wooden spear. The plunger-case a is secured to the top of an H-piece, consisting of a short pipe J, and the suction clack piece K, united by the waterway or

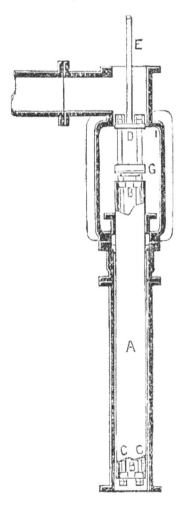

Figure 82 Hollow Ram Pump

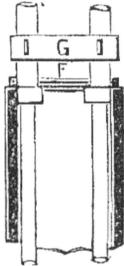

Figure 83 Top of Plunger

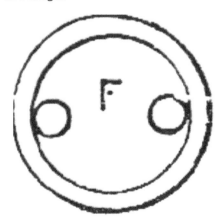

Figure 84 Plan of Delivery Valve

throat I. The suction clack piece is bored at e to receive the suction clack; h is the wind bore, or suction pipe, bolted to the underneath flange of the H-piece; and the pipe J is secured to timber built into the sides of the shaft. On the top of the H-piece is secured the delivery clack-piece, which is bored at F to receive the delivery clack. g is the delivery pipe or pump trees, so called because originally they were made from logs of trees, bored out and strengthened by wrought-iron hoops.

b. A single-acting hollow plunger-pump is illustrated in Figures 82, 83, and. 84. Figure 82 is a sectional elevation; Figure 83 gives a view of the top of the plunger; and Figure 84 shows the plan of the delivery-valve. The plunger-case is similar to the ordinary one, and is fitted at the top with a stuffing box and gland. To the top of the stuffing box is secured a delivery clack- piece, furnished with a door for access to the delivery clack; and to the top of this clack-piece is bolted a cover, provided with a gland and stuffing-box, for the pump-rod (not illustrated). The plunger consists of a hollow pipe a without flanges, turned all over perfectly true and parallel; or, at least, it ought to be so. It is furnished at top and bottom in the inside with two lugs b, through which pass two wrought- iron rods C, provided at the bottom ends with double nuts, and at top secured to a crosspiece d. forged on the main pump rod E and secured by double nuts. The delivery valve F in this example of hollow ram or plunger-pump is simply a round disc sliding on the two rods C, the disc being faced with leather. A cross-piece G, secured to the plunger-rods C by colters, limits the lift of the valve.

c. Double-acting plunger pumps are divided into those that have no packing, those that have internal packing, and those that are externally packed.

It is thought that those without packing are the oldest, and were invented by the late Mr. Henry Worthington. One of these pumps is illustrated in Figures 85 and 86, Figure 85 being a sectional elevation and Figure 86 a cross-section, B is the plunger, which works through a metallic gland or bush C without any packing; D, D are the suction valves and E, E the delivery valves, consisting each of an indiarubber disc, rising on a brass spindle, with a guard at the top, and falling upon a circular grid or seat, perforated with holes. In the original pump of this type a few holes H, H were drilled, which had the effect of opening a communication between the two ends of the pump barrel at the end of each stroke, thus giving the water, as it were, a partial elasticity, and allowing it to continue its forward motion by flowing through the plunger the moment that the plunger becomes stationary. This enables the plunger to begin its return stroke without any shock or concussion. O is an air vessel on the delivery side, or what is called a pressure air vessel, and P is a suction air

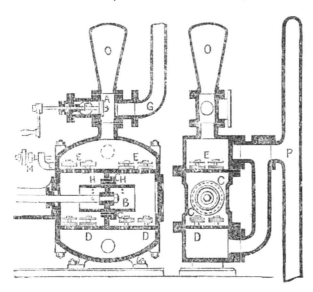

Figure 85 Double-acting Plunger Pump, without Packing

Figure 86 Double-acting Plunger Pump, without Packing

vessel, or, more correctly, a vacuum vessel. A is a double-seated valve, by means of which the water can be delivered either through the pipe F or G.

An internally packed double-acting plunger pump is shown in sectional elevation in Figure 87, and an end view, with the back cover, or bonnet, removed, is shown in Figure 88. This pump is used for moderate lifts in mines and collieries; it is a single solid plunger, worked as a double-acting plunger. This is done by providing the barrel of the pump at or near the centre with a stuffing-box for the plunger to work through, and placing each end of the working barrel in connection with a set of suction and delivery valve-boxes. The valve-boxes may be cast on, as illustrated, or made separate, and bolted to the pump body, as may be found most convenient. In this example A is a

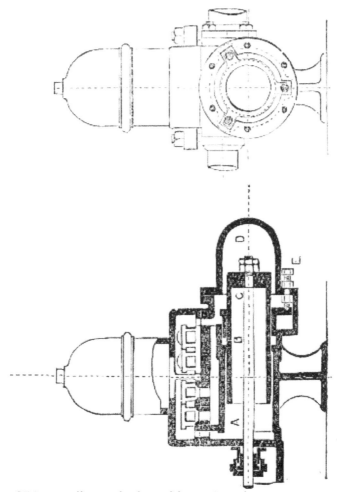

Figure 87 Internally-packed Double-acting Plunger Pump

Figure 88 Internally-packed Double-acting Plunger Pump

single barrel, which is provided near the centre with a stuffing-box B for the plunger C to work through, and each end of the barrel is in communication with a set of suction and delivery valves. The valve-box cover has an air vessel cast on it, and a communication is made between the delivery-valve box and the air vessel, D is a cover, or bonnet, which covers the back end of the plunger, and in which the plunger works; at E are three screws, fitted with check or back nuts, for tightening the internal gland.

Externally, packed pumps are also in use. When pumping against heavy pressures, it is desirable to expose the stuffing-box and gland round the plunger of the pump, so as to make them more accessible, and enable one to see if they are packed sufficiently tight to prevent leakage. For cases of this kind the barrel of the pump is usually cast in two pieces, and joined together by a connecting piece, or distance piece and bolts; or, instead of casting the barrels separ-

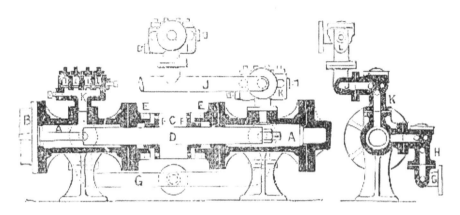

Figure 89 Externally-packed Double-acting Plunger Pump for Moderately Heavy Pressure

Figure 90 Externally-packed Double-acting Plunger Pump for Moderately Heavy Pressure

ately, they can be cast together by a framework joining their two inner ends, and thereby forming the two barrels and framework in one solid piece. The inner ends of the barrels are provided with stuffing- boxes and glands, through which the plunger works. The valve-boxes may be cast to the barrels, or separately made and bolted to them. The barrels of the pump are made so that the inlet and discharge valves may be placed on either side of the pump. Pumps of this kind may be constructed to work in a horizontal position, as illustrated, or in a vertical position. The piston-rod of the engine is attached to one end of the plunger, and works through a stuffing-box and gland in a cover on one of the outer ends of the pump barrel.

Figure 89 is a longitudinal section through the pump barrel and one of the delivery-valve boxes; and Figure 90 a cross-section through the pump barrel and valve- boxes. a, a are the pump barrels; B, the distance- piece between the steam cylinder and pump. The

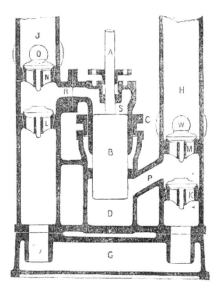

Figure 91 Externally-packed Double-acting Plunger Pump, with one Gland only

two pump barrels are joined by the flange distance- piece C. The plunger D works in the two pump barrels, and is packed in its middle by the stuffing- boxes E, E and glands F, F; G, the inlet or suction- pipe, connects the pump barrels by the valve-boxes H, H; J is the delivery pipe, attached to the top of the pump barrels by the delivery-valve boxes K, K. This pipe has a valve-box L to serve as a retaining valve.

Another externally-packed double-acting plunger pump is shown in Figure 91. This differs from the previous one in that only one stuffing-box and gland is used for packing the plunger. The pump consists of a plunger B actuated by the pump-rod A. The plunger works through a stuffing-box and gland C, which are placed in the upper end of the lower or

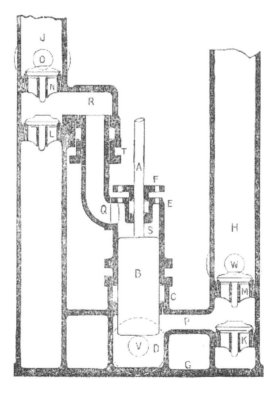

Figure 92 Externally-packed Double-acting Plunger Pump, one Working Barrel acting as Gland

bottom cylinder or working barrel D; S is the top working barrel. It will easily be seen that when the gland C is tightened the packing will prevent any leakage from either of the working barrels D and S. G is the suction chamber; K and L the suction valves; M and N the delivery valves; O and W the two delivery pipes (O being the delivery pipe from the top working barrel S and W from the bottom working barrel D); P is the passage admitting the water or other liquid to the barrel D, and R the passage admitting to the top barrel S; J and H are two air vessels on the delivery side of the pump.

There is still another method of arranging the packing for externally-packed double-acting plunger- pumps. In Figure 92 A is the pump-rod, to which is connected the plunger B, worked up and down or to and fro by any arrangement of motive power. The

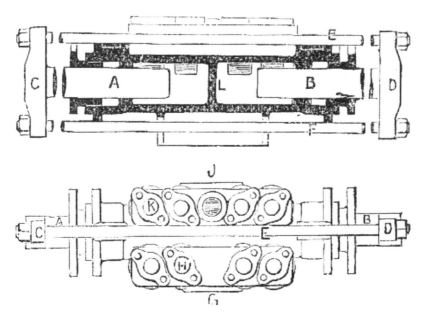

Figure 93 Externally-packed Double-acting Plunger Pump, with Two Plungers

Figure 94 Externally-packed Double-acting Plunger Pump, with Two Plungers

plunger B works through a stuffing-box C placed in the upper end of the working barrel D. The upper working barrel S forms the gland. This working barrel is provided with a cover or top E and stuffing-box and gland F. The two working barrels D and s fit together, as illustrated, and are bolted together. The barrel D is in one casting with the foundation-plate G and the columns H and J, the lower ends of which are provided with seatings for the inlet or suction valves K and L, and the delivery valves M and N, while the upper parts of the columns form delivery air vessels, to keep up a constant flow in the delivery pipes O and W. The passage p connects the pump barrel D with the right-hand column H and the bent pipe Q; and passage R connects the upper portion S of the pump barrel to the left-hand column J. The pipe R passes through a stuffing-box T in the upper part of the column J.

Externally-packed double-acting plunger pumps with two plungers will now be noted. Pumps with only one plunger are very useful for low and moderate pressures; but when dealing with heavy pressures, it is found that the total pressure is greater on one side than the other, on account of the area of the piston or pump-rod side being deducted from the area of the ram on one side. The pump illustrated in sectional

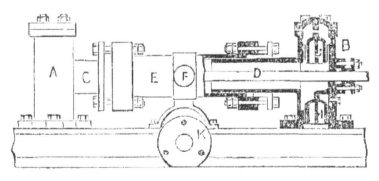

Figure 95 Double-acting Plunger Pump with Fixed Plungers

elevation (Figure 93) and plan (Figure 94) obviates this fault, because there is no piston-rod on either side. It was originally manufactured at the Knowles Steam Pump Works, Warren, Mass., in the year 1867. In this case the areas exposed to the pressure are perfectly balanced, the two rams being of the same diameter; hence a uniform load is presented to the steam cylinder or pumping gear at each stroke of the plungers, A and B are the two plungers, each provided with a cross-head C and D, and the two cross-heads are coupled together by means of the two rods E and F. The pump body is cast in one, having a diaphragm or divisional plate L in the middle of its length, so as to form two working barrels, into which the plungers A and B work. One of the cross-heads is coupled to the pump-rod. The plungers work through glands and stuffing-boxes, which are clearly illustrated, G is the inlet branch, H the suction valves, J the delivery branch, and K the delivery valves. The top rod E works on guides bushed with white metal.

The externally-packed double-acting plunger pump shown by Figure 95 differs from all the others, inasmuch as that, instead of the plunger having a reciprocating motion, there are two fixed plungers, and a to-and-fro motion is given to the working barrel. This pump was invented by Mr. Thomas Dunn, of Manchester, and was exhibited in the London Exhibition, 1862. Figure 95 is half elevation and half sectional elevation, A and B are the two valve-boxes, firmly secured to the bed-plate by means of bolts and nuts; c and d are the two plungers, cast in one with the two valve-boxes, and turned true and parallel. On these plungers, or rams, the working barrel E is moved backwards and forwards by means of the cross-head pin F and a connecting-rod. h is the suction valve, J the delivery valve, and K the suction pipe. The delivery pipe (not shown in the illustration) can be arranged at any convenient place above the delivery valve J.

D. Bucket and plunger pumps are a combination of a bucket and a plunger pump, the plunger being one- half the area of the bucket. Some authorities attribute the invention of this pump to Smeaton, the famous English hydraulic engineer, who was born in the year 1724, and died 1792; but the " Theatrum Machinarium" (written by Jacob Leupold, 1674-1727) contains a description of one type of this pump, in which the diameters of the two pistons are different, " usually in the proportion that the area of the larger one is double that of the lesser one, being moved by the same rod. "

The bucket and plunger pump is frequently termed differential pump, because its action depends upon the different areas of the bucket and the

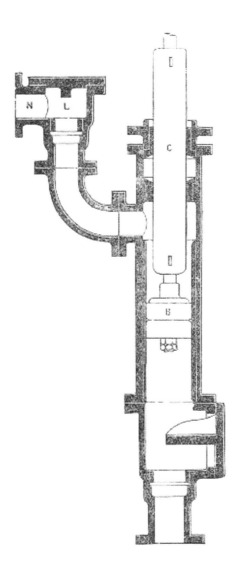

Figure 96 Vertical Bucket and Plunger Pump

plunger. It is illustrated in sectional elevation by Figure 96.

When the bucket B is raised, it sucks—or rather, draws—a column of water equal to its area, its length being the same as the length of the stroke; at the return stroke, when the bucket is travelling downwards, the water passes through the valve in the bucket to the upper part. The ram or plunger C,

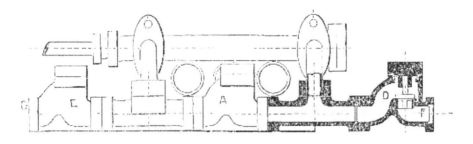

Figure 97 Horizontal Piston and Plunger Pump

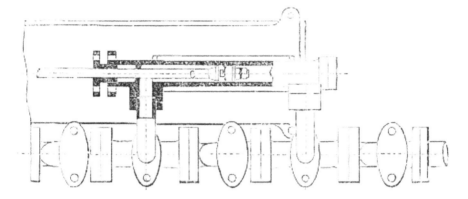

Figure 98 Horizontal Piston and Plunger Pump

being one-half the area of the bucket B, leaves an annular space round the plunger equal to half the area of the entire space of the suction side; therefore, it will be found that half the quantity of the water is raised into the rising main N through the delivery valve L, the other half portion remaining in the working barrel above the bucket. When the bucket again moves upwards, the remaining half of the water is lifted into the delivery pipe N. From this it will be seen that half the quantity is delivered at the up and half at the down stroke; consequently the pump is double-acting, but delivers only the same quantity as a single-acting pump of the same diameter as the

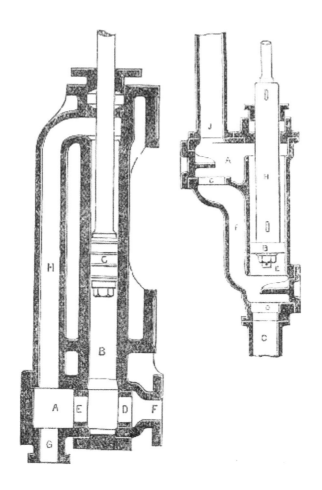

Figure 99 Horizontal Piston and Plunger Pump

Figure 100 Horizontal Piston and Plunger Pump

bucket, or as a double-acting pump of the same diameter as the plunger.

e. The piston and plunger pump is another pump with single-acting suction and double-acting delivery; but in this case the piston is solid, so that the water has to pass through a passage outside the barrel instead of through the piston. Some engineers prefer this latter arrangement, because there is less, friction in the passage than through the bucket. The pump is of the horizontal type. A pump of this description is shown in part sectional elevation (Figure 97) and part sectional plan (Figure 98). a is the delivery-valve box; B the piston; C the plunger, or piston-rod in most cases; D is the suction valve and box; E a retaining, or back-pressure, valve-box; F the suction or inlet pipe; G the delivery, or outlet pipe. The action is exactly similar to that of the bucket and plunger pump.

Lord William Armstrong said, in a paper read by him before the Institute of Mechanical Engineers in the year 1858, that " this type of pump was suggested to me about the year 1851 by Mr. Henry Thompson, my late intelligent foreman. "

A modification of the above design is illustrated in Figure 99, which is a sectional elevation of another horizontal pump. The advantage of this design over the previous one is that the delivery-valve box a is arranged above the working barrel B, so that there is no air-lodge formed; secondly, the clearance space between the piston C and the suction and delivery valves D and E is reduced to a minimum, which greatly improves the so-called suction power of the pump. F is the suction inlet; G the delivery branch; and H a passage leading from the back of the piston C to the ram or plunger end of the pump.

A piston and plunger pump of the vertical type is illustrated in sectional elevation at Figure 100. B is the piston, which is solid, without any valve; H is the plunger, or ram, secured by a cotter to the pump- rod; D is a recess for receiving the suction-valve seat; F is a passage at the side of the working barrel, communicating with the delivery valve, which is secured into the space G; a is the delivery-valve box; and J the delivery pipe, or rising main.

f. Plunger and Plunger, and ram and ram, refer to the same type of pump, which was patented by Mr. Thomas Heaton in the year 1844; it is, in reality, a combination of a single-acting ram and a hollow ram pump. A, in Figure 101, is the hollow ram working through a stuffing-box and gland; B is a ram of one-half the area of the hollow ram A; C is the working barrel for the hollow ram; and the ram B is working in the pump-trees, or delivery-pipe, H; D is the suction pipe; E the bored end for receiving the suction clack; E the suction clack piece, or valve-box. The hollow plunger A is fitted with two valves or delivery clacks J and K, so that if one should fail at a critical moment the other would in all probability be in working order. One of these valves J is fitted at the bottom of the ram, and is accessible through the suction clack piece F; the other clack K is fitted at the top of the ram A, and is accessible through the delivery clack piece G. The action will be readily understood, as it is the same in the bucket and plunger pump. The plunger B can be made of wood, in the shape of a spear, if cut to the proper sectional area. The above pump is known also as the double- acting hollow ram pump.

The solid ram and ram pump is similar to the piston and plunger pump, the difference being that the piston is prolonged to a ram. This is illustrated in Figure 102. a is the large ram, or plunger; H the smal. one, having an area one-half that of the area of the ram A; J is the piston-rod or pump-rod; B is the suction valve, and e delivery valve; M is the suction branch, and F delivery branch; G is a passage leading from the suction side to the delivery side of the large plunger A; K is the water-tight packing, in which the large ram a works. In this case the packing consists of two U leathers, but in some instances it is made in the shape of an ordinary stuffing-box and gland, L is the stuffing-box and gland for the small ram H, which also consists of leather packing.

With regard to the advantages of ram pumps in general, there is less friction against the inside of the working barrel in pumps where the ram works through a stuffing-box than in the bucket or piston pumps. When the water is impregnated with sand, grit, mud, and other foreign matter—as is generally the case in collieries and mines—the wear and tear of the plunger in horizontal ram pumps are prevented by the space underneath it forming a receptacle for the solid matter, which is not the case in the bucket or piston pumps. The cost of re-packing the ram or plunger is only nominal; but the cup leathers or other leathering for the pistons and buckets are a great expense. In cases where the ram is packed from the outside, leaks are immediately detected, and the gland tightened in a few minutes, while the pump is working, and the stuffing-box re-packed in a very short time; whereas in a piston pump or bucket pump a leak can go on for a long time without being noticed. The- ram can be lubricated, and the corroding action of acidulated water greatly reduced.

There is no such thing yet invented as a perfect pump—that is, one suitable for all purposes. Some designs have one advantage, others another. For instance, Figure 78 (p. 69) shows a single-acting plunger pump in which the flow of the water through the suction-pipe is direct into the plunger-case; hence there is little frictional resistance to the water passing through the valve. On the down stroke the friction is greater. The resistance on the suction side must be as little as possible, as the so-called suction depends upon the pressure of the atmosphere on the surface of the well or other source, but on the delivery side it matters less, as the limit on that side is only regulated by the motive power. The fault of this design is that the water passing straight up delivers all the air contained in it—the amount of which in some cases is very considerable—into the plunger-case; so that at times the whole plunger-

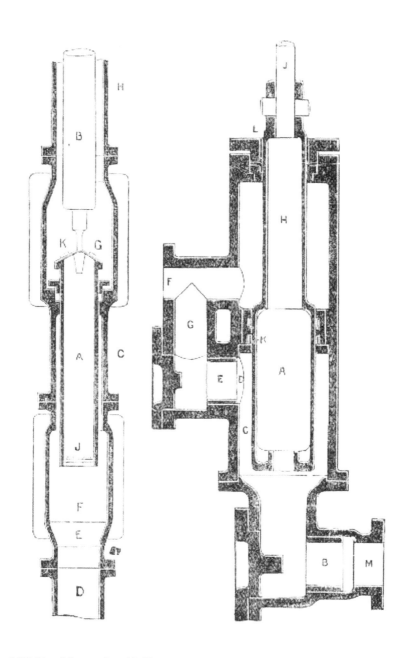

Figure 101 Double-acting Hollow Ram Pump

Figure 102 Horizontal Plunger and Plunger Pump

case may be full of air, which is compressed and expanded at every stroke, and very little, if any, water enters the pump. To remove the air, the cock is fixed at the top of the plunger-case, and this cock has to be opened at intervals. In some cases the cock is omitted, and the packing in the stuffing-box allowed to be just slack enough to let the air escape.

Figure 79 shows the Trevithick ram pump, in which the flow of the suction is the same as in Figure 78; but the delivery clack being at the top of the plunger case, the air admitted with the water will pass straight out of the delivery valve and up the rising main. However, there is one great defect in Figure 79 not present in Figure 78—namely, that the plunger case must be made twice the area of the plunger, to allow the water to pass round it; this produces a very large clearance space between the suction and delivery valves and the plunger, which deteriorates the vacuum produced by the pump. That, of course, in pumps having a very short suction matters little, but in cases when the height of suction is great it is a grave error.

Figure 85 illustrates a double-acting ram pump without any packing, showing a great amount of clearance space, which is detrimental to the suction power; in fact, some of these pumps have such an amount of clearance space that they will not fetch water more than 12 ft. —which is very little indeed. A double-acting pump of this description ought to be able to draw the water at least from 18 to 20 ft., and in many instances well-designed pumps have fetched water from a depth of 25 feet. One advantage of this pump is that the diameter of the plunger can easily be changed for a larger or a smaller one by changing the bush C for one of suitable size; but the less the diameter of the plunger the larger the clearance space. Figure 87 shows an arrangement where the clearance space is small, but the air-lodges very bad, all the valves, both suction and delivery, being on the same plan. The air above the suction valves cannot be cleared without an air- cock being fitted to the cover above them. Another objection to this pump is the internal gland and stuffing-box, although in this case the stud E for tightening the gland is outside, so that it can be screwed up without stopping the pump or removing the bonnet or cover D, but it must be stopped for repacking the stuffing-box.

An excellent arrangement of valves and pipes is the one shown in Figures 89 and 90. Here the suction is at the bottom below the centre of the working barrel, so that the air entering the suction pipe will pass through, up into the rising main, without the least chance of being stopped on its way thence.

Two good samples of vertical double-acting ram pumps are those illustrated in Figures 91 and 92. The passage for the air is quite straight, so that very little air is likely to get into the barrels S and D; and should any get in it will be returned directly, because the passages P, R, and Q are at the top of their respective working barrels. One objection exists in Figure 92—namely, that there are three glands and stuffing-boxes, E, F, T, and C, to attend to, whereas Figure 91 has only two, namely, one for the plunger and one for the pump-rod.

Next in order is the double-ram pump, illustrated in Figures 93 and 94. It is a very excellent design; the passages for the air are clear and direct, and the clearances are reduced to a minimum. However, it is more expensive, and takes up rather more room than the usual designs of double-acting ram pumps. For very heavy pressure it is, without a doubt, the best arrangement possible to adopt.

The pump shown by Figure 95 is more remarkable for its curious construction than for any advantages it possesses. The wear and tear and friction produced by the heavy working barrel moving backwards and forwards must be greater than when the plunger is moved; hence it is more difficult to keep the glands and stuffing-boxes tight, and the clearance spaces are a great deal larger than in the ordinary double-acting plunger pumps. The arrangement as regards air is very good, and this is the only redeeming feature in the design.

The greatest advantages claimed for the bucket and plunger pump are that it is double acting, with only two valves, and also that the work done is equal both at the up and down stroke, which latter has been previously described. In Figure 96 it will be seen that the arrangement of the passages is perfectly straight; hence the air has a free passage, and the clearance space is reduced to the utmost limit. The objection to pumps of this class is that they have the friction both of the bucket and the plunger. The suction pipe must be made at least one-half the area of the bucket, and the delivery pipe should be about one-half the area of the plunger. Another objection is the contracted area of water-way through the bucket.

One of the worst designs of piston and plunger pump is the one illustrated in Figures 97 and 98. The suction and delivery valves, being below the working barrel, form very bad air-lodges, and the clearance space is great; in fact, in some cases it is two, or even three, times the displacement of the piston. Figure 99 shows just a contrary case; here the delivery valve is directly above the suction valve, so that the air will pass straight out; the clearance space is reduced to its utmost limit, and there is a free and easy flow for the water, instead of the distorted and curved passages in Figures 97 and 98.

The last objection to bucket and plunger pump —namely, contracted area of the water-way through the bucket—is done away with in the piston and plunger pump (Figure 100), the piston being solid and the water being passed up the passage F to the delivery valve G; but this arrangement increases the clearance space.

It will be seen that in nearly every pump yet invented, or in every different design of pump, when one advantage is brought forward, one or more disadvantages frequently present themselves. The flow of the air is good in Figure 100, but the friction is great, there being both a piston with packing and the gland and stuffing-box for the ram or plunger.

The ram and ram pumps, illustrated in Figures 101 and 102, may now be discussed. The first (Figure 101) is the hollow ram type, which has the same advantages and disadvantages as the single-acting hollow ram pump; but it has the additional disadvantage that it has the friction of the two rams in their respective stuffing-boxes. It has, however, only two valves, although Mr. Heaton, the inventor, in this instance, prefers to arrange double sets of valves for safety. This is certainly a good plan in cases where the level of the water is liable to sudden rising, which is frequently the case in wells, and covering the clack-door F. The internal gland and stuffing- box for the plunger a is also objectionable. The clearance space round the plunger A, in Figure 102, is large, because the delivery valve is placed over the barrel instead of over the suction valve, otherwise the space is reduced to a minimum; however, the reduction has caused an objection—namely, that before the plunger can be removed, the joint which connects the suction pipe to the suction valve box B must be broken and the valve box removed. The internal packing for the plunger a is objectionable; in this case it consists of two hydraulic U-leathers placed back to back.

There are some special points to be attended to in designing ram or plunger pumps. The point which must have the first and most earnest consideration, and which is mostly overlooked altogether, is air-lodges. As will be clearly seen in the piston and plunger pump (Figures 97 and 98), a better and plainer example it is impossible to find, because the air-lodges are as bad and numerous as possible,, both on the suction and delivery side of the pump. As an example of good arrangement, note Figure 99.

The next important point to bear in mind is that- the clearance space should be as small as possible, because the vacuum is impaired when the space is- large; hence the height from which the pump can fetch the water is reduced in the same proportion as the clearance space is increased. The clearance space in a well-designed pump is never larger than the displacement of the plunger. Easy passages for the flow of the water or other liquid through the pump and valves should always be arranged.

Sharp corners should always be avoided; long, easy curves should be used where possible; the radius of bends ought never to be less than the diameter of the pipe—twice the diameter is better; elbows should never be used on any consideration.

The length of the stroke should be as great as possible, because the reversal of the pump always causes, even in the best designed plunger pump, an unavoidable shock at the end of each stroke; besides, the valves having to open and close fewer times, the piston speed being the same, the wear and tear of the valves are reduced, and the amount of water lost by " slip " (that is, water that returns into the suction pipe at the reversal, before the valves have had time to close) is less.

The plunger should always enter into the bored part of the working barrel at every stroke, so as to prevent any ridges forming on the plunger or in the working barrel.

Doors for easy access to the valves should always be provided for examination and repairs. The valvebox doors on large pumps which are hung vertically should have a lug on the top edge, or on the face, to lift it by when it has to be removed. Horizontal covers or doors should be furnished with a boss and an eye-bolt in the centre, for the same purpose.

Internal packings should be avoided when possible, especially as the liquid to be raised is impregnated with sand, grit, or other impurities.

The plunger should be well-rounded at the end; in fact, it is best to make the end hemispherical in shape, to reduce the shocks at the return stroke.

The water-way through the bucket should be made as large as possible, so as to reduce the friction. When the plunger is actuated by means of a crank and connecting-rod, and no guide is employed, the bored part of the working barrel should be made extra long, and the connecting-rod passed down in the plunger as deep as the clearance of the connecting- rod will allow, so as to reduce to the utmost degree the tilting action produced by its angularity.

In the bucket and plunger and piston and plunger pumps, the piston or bucket, as the case may be, should be exactly double the area of the ram or plunger, or else shocks will be produced in the delivery pipe by the two speeds at which the water is delivered meeting one another.

The plungers should never be draw-filed—as is sometimes specified in large pumping plants for water-works and sewage-works pumps. They should be turned with a well-defined cut, and will soon work to a smooth surface, which will last a long time.

No studs should be used wherever bolts can be put in, as much annoyance, loss of time, and often a great loss of money, are caused by a stud breaking, as the old stud must be generally drilled out before a new one can be put in; bolts, as a rule, are also cheaper.

Uniformity of metal all through the pumps, to equalise the contraction in cooling the casting, should be well considered.

Sudden enlargements and contractions in the pump passages should be avoided; but if any alterations in size or shape of the passages are imperative, they should be made gradually.

CHAPTER VI

MAKING BUCKET AND PLUNGER PUMPS.

The object of this chapter is to give instructions on making a simple and efficient pump that may be constructed by any person handy with tools, but without a lathe.

The materials required are the ordinary wrought- iron pipe, flange and tee-pieces, drawn brass tube for the barrel, and thick sheet brass for the buckets.

The important points in the design of small power pumps are: the water in the rising main must have a constant flow, and there should be few working parts.

The type most nearly approaching these points is the bucket and plunger pump, this type having all the advantage of the double-acting pump, the water in the rising main always being in motion, with the simplicity of the single-acting pump, having only two valves.

As the power and other conditions vary, it is proposed to give a general description and drawings of the pump, with tables giving the dimensions for the different parts for pumps varying from 1 1/4 in. to 3 in. in diameter. The dimension of each part is found from the table under the letter corresponding to that on Figures 103 and 104. A stroke of 4 in. is adopted for all diameters given in the tables. The length of the working barrel will be the same for each, but the total height will increase slightly with the diameter, this being due to the difference in length of the iron fee-piece. A' is the working barrel of drawn brass tube 7 in. in length, soldered water-tight to the wrought-iron flange B. This flange forms the foot for bolting the pump to the plank, as well as the joint for the suction pipe. An iron tee-piece C is also soldered to the brass barrel, the tee having half the area of the barrel.

The construction of the bucket and valve is shown

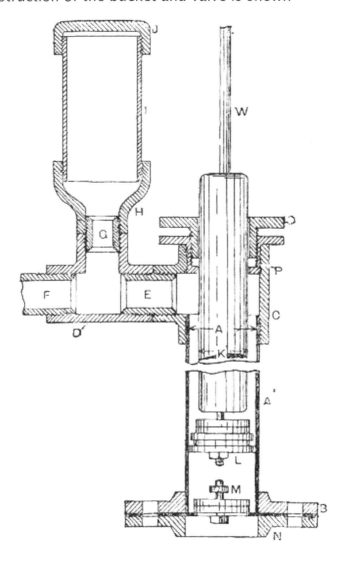

Figure 103 Section of Bucket and Plunger Pump

in detail in Figures 104 and 105. The stuffing-box is constructed of short lengths of screwed pipe and back nuts. At P (Figure 103) is a short length of pipe fitting the tee-piece, with a back nut at the top to form a flange, the nut being firmly fixed to the pipe by riveting over after being screwed down flush with the end of the pipe. At the lower end is soldered a brass ring sufficiently large to allow the plunger K to pass with a good fit. Above this ring is the space for the packing. This piece can be unscrewed to allow the plunger and bucket to be removed for renewing the cup leather. The gland O is another piece of pipe with a back nut. There being no screw on the inside of P, the gland is tightened up on the packing by tee-headed bolts and nuts, the bolts fitting slots cut in the back nuts. Two bolts will be sufficient for the 1 1/4-in. and 1 1/2-in. pumps; larger sizes will require either three or four bolts equally spaced round the back nuts. The dellvery pipe is taken off the tee with a nipple E, this having half the area of the working barrel. At D is another tee-piece for the iron air vessel connection; G is a nipple making connection between D and the reducing socket H. The air vessel I with cover J has the same diameter as the working barrel. F is a nipple making connection with the delivery pipe. The lower valve M is made of sheet brass having the same dimensions as the buckets, except the valve seat, which is of the same diameter as the flanges B and N. Thin sheet rubber or oiled brown paper is placed between the flanges and valve

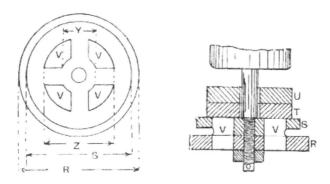

Figure 104 Waterways in Buckets and Bottom Valve

Figure 105 Section of Bucket

seat to make an air-tight joint. The waterways V are cut through the brass plate as shown in Figure 104.

The buckets are of sheet brass 1/4in. thick. All the bucket parts and lower valves are of the same thickness, the diameters and areas of waterways varying.

The bucket (Figure 105) is constructed as follows. Three discs R, S, and U are of sheet brass, R being a good fit in the barrel. Through the centre is drilled a hole for the rod W, and four waterways are cut as shown at V. A recess in. wide and 1/4 in. deep is cut on the edge of disc S to receive the bottom of the cup leather. The centre hole is tapped to screw on the rod W and fit well against the shoulder of the rod. U is a disc fitting loosely over the rod, and T is a leather or rubber disc sliding over the rod. These

Figure 106 Set-screw to Prevent Tee Turning on Nipples

discs form the valve, and should have a lift of from 3/16 in. to 1/4 in.

The bucket is put together in the following order: First the disc U, then the rubber or leather T, and below this the body and valve seat s with the cup leather. The whole is made secure by the lower disc R being brought up tight against the cup leather by screwing up the nut on the pump rod. With the smaller pumps it will be advisable to have a back nut to prevent the bucket working loose, but with the larger pumps a split pin may be used. Care must be taken when putting the last disc on that the waterways coincide with those in the disc S. The lower valve is constructed in a similar manner, but only two brass discs are required, the valve and seat.

The valve spindle has a head to check the lift of the valve. The spindle is bolted to the seat by a nut on the under side.

The plunger is a brass tube with the two ends stopped with a 1/4-in. brass disc soldered in. Through the centre passes the rod W (Figure 103). The ends of the rod are threaded below the shoulder. To prevent the tee-pieces turning on the nipples E and F, holes may be drilled through them for a small set-screw as shown in Figure 106.

The suction pipe is connected to the lower flange N with a reducing socket and two nipples, and should have an area of three-quarters that of the barrel.

The following table gives dimensions of the parts. The figures for the flanges, tees, reducing sockets, and caps are those known in the trade to fit tubes of the diameters given. Water tubing should be used for high lifts.

TABLE I

Size.	A.	C.	D.	E & F.	G.	H.	B,I,J,N.	W.
in.	in.	in. in.	in. in.	in.	in.	in. in.	in.	in.
$1\frac{1}{4}$	1	$1\frac{1}{4}$ to 1	1 to $\frac{1}{2}$	1	$\frac{1}{2}$	$1\frac{1}{4}$ to $\frac{1}{2}$	$1\frac{1}{4}$	$\frac{5}{16}$
$1\frac{1}{2}$	$1\frac{1}{16}$	$1\frac{1}{2}$ to $1\frac{1}{4}$	$1\frac{1}{4}$ to $\frac{1}{2}$	$1\frac{1}{4}$	$\frac{3}{4}$	$1\frac{1}{2}$ to $\frac{3}{4}$	$1\frac{1}{4}$	$\frac{3}{8}$
2	$1\frac{1}{2}$	2 to $1\frac{1}{2}$	$1\frac{1}{2}$ to $\frac{3}{4}$	$1\frac{1}{2}$	$\frac{3}{4}$	2 to $\frac{3}{4}$	2	$\frac{5}{8}$
$2\frac{1}{2}$	$1\frac{3}{4}$	$2\frac{1}{2}$ to $1\frac{3}{4}$	$1\frac{3}{4}$ to $\frac{3}{4}$	$1\frac{1}{2}$	$\frac{3}{4}$	$2\frac{1}{2}$ to $\frac{3}{4}$	$2\frac{1}{2}$	$\frac{1}{2}$
3	2	3 to 2	2 to 1	2	1	3 to 1	3	$\frac{1}{2}$

Instead of webs for the 11/4-in. and 11/2-in. pumps, holes should be drilled, leaving in. between them to form the waterways. These in the larger pumps must be cut with 1/4-in. webs as shown in Figure 104.

The capacity of the barrels for each stroke, allowing 10 per cent, for slip, is: —1 1/4 in., .014793 gal.; 1 1/2 in., .021303 gal.; 2 in., .037872 gal.; 2 1/2 in., .059175 gal.; and 3 in., .086112 gal.

The size of pump for a given power may be found in the following way. Multiply the gallons required to be lifted per minute by 10 and by the vertical height in feet, when the product will be the theoretical

TABLE II

Size.	*U. & T.*	*S.*	*R.*	*Y.*	*Z.*
in.	*in.*	*in.*	*in.*	*in.*	*in.*
$1\frac{1}{4}$	$\frac{15}{16}$	$1\frac{1}{16}$	$1\frac{1}{4}$	$\frac{3}{8}$	$\frac{13}{16}$
$1\frac{1}{2}$	$1\frac{1}{4}$	$1\frac{1}{4}$	$1\frac{1}{2}$	$\frac{9}{16}$	1
2	$1\frac{5}{8}$	$1\frac{3}{4}$	2	$\frac{9}{16}$	$1\frac{3}{8}$
$2\frac{1}{2}$	2	$2\frac{1}{4}$	$2\frac{1}{2}$	$\frac{3}{4}$	$1\frac{11}{16}$
3	$2\frac{3}{8}$	$2\frac{3}{4}$	3	$\frac{7}{8}$	2

power; then with small pumps double this to allow for the friction. The quantity of water lifted will depend on the number of strokes.

The diameter of a pump suitable for a given power and lift is found as follows: —Assume that a water- wheel develops 1,800 foot-lb. and makes twenty revolutions per minute, and that the water is to be raised 50 ft. The water lifted in gallons per minute is found by dividing the foot-lb. by 2 and by 10 (the weight of one gallon in lb.) and by the height in feet. It will therefore be

$\frac{180}{2X50X10}$ = 1.8 gal. a minute. From this the diameter is found by dividing the gallons per minute by the number of strokes and consulting Table III. The diameter will be 1.8 ÷ 20

= .09; the nearest to this in the list is .086112, opposite which will be found 3; therefore the diameter to employ will be 3 in.

TABLE III

Diameter.			Gal. per stroke.
$1\frac{1}{4}$ in.	..	..	·014793
$1\frac{1}{2}$,,	..	..	·021303
2 ,,	..	..	·037872
$2\frac{1}{2}$,,	..	..	·059175
3 ,,	..	..	·086112

CHAPTER VII

CONSTRUCTION OF PLUMBER'S FORCE PUMP

The illustrations in this chapter show a force pump that can be very easily and cheaply made. The pump (Figure 107) is made up of plumbers' brass and copper fittings. The total height is 1 ft. 9 5/8 in. over all.

The first requirement is a piece of 2-in. copper or brass tube, 14 1/4 in. long, clean inside and free from scale, etc. The tube should be thoroughly scoured inside with coarse and fine emery cloth, to ensure a smooth surface. It should be perfectly square on the ends, and then tinned at each end, inside and out, 1/2 in. up.

Figure 108 shows the bush-piece required. This is tinned on the male end and on the flange, and then soldered to the end of the 2-in. tube. A 1 3/4-in. hole is then cut in the side of the tube, as near as possible to the bush-piece. This hole is tinned 1/2 in. all round, and a similar bush-piece to that already described is tinned and soldered in, no part of the bush-piece being allowed to project inside the tube, and all surplus metal being cut off before soldering in.

A plumbers' brass 2-in. screw cap is tinned and soldered on to the other end of the tube, the square nut usually on the cap being cut off, and a 7/8-in. hole bored through the centre as shown in Figure 109.

The next thing required is the top part of a 1/2-in. brass screw-down bib-cock. The crutch of the cock being cut off and the rod screwed out, the part left can be used as a guide for the suction rod as shown in Figure 109. The flange is tinned and soldered into the hole on top of the screw cap (see Figure 109).

The suction rod is of 3/8-in. brass or copper rod, 15 in, long, with a 1/4-in. gas thread on one end, 1/2 in.

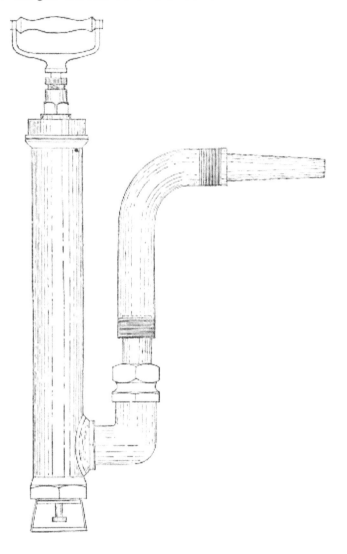

Figure 107 Plumber's Force Pump

long, on which to screw the handle. (A valve closet handle is shown in Figure 107.) Then a 1/4-in. gas thread, 1 in. long, is formed on the other end, which is also provided with two 1/4-in. gas back-nuts and two 1 5/8-in. circular discs of brass, with a 3/8-in. hole through their centres; while a 2-in. cup leather, with a 3/8-in. Hole through the centre, and fitted up as shown in section (Figure 110), care being taken not to have the rod too long.

The suction valve seen in section (Figure 111) is made up with a 1-in. male to 3/4-in. internal thread bush-piece, to screw into the 1 1/2-in. to 1-in. bush-piece.

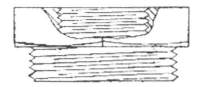

Figure 108 Bush Piece for Plumber's Force Pump

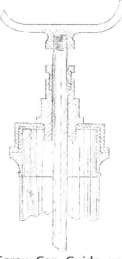

Figure 109 Section of Screw Cap, Guide, and Handle for Plumber's Force Pump

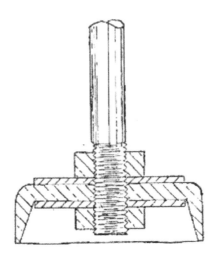

Figure 110 Suction Rod and Section of Cup Leather, etc., for Plumber's Force Pump

A 1-in. jumper (Figure 112), out of a 1-in. brass screw-down stopcock, has its edge filed or turned off (see Figures 111 and 112). A thread is put on the long end, and the nut taken off the short end is screwed on to the long end as shown in Figure 111. The short end of the jumper is left on to receive the hand brace for grinding in to the top of the 1-in. bush-piece, which should be tapered inwards for the seat of the

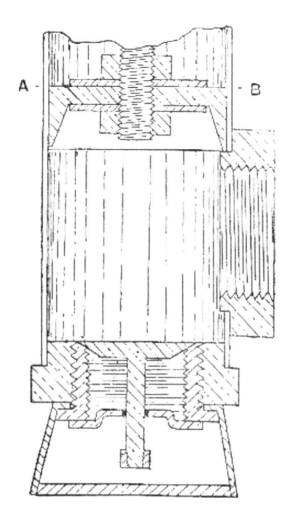

Figure 111 Section of Bush Pieces, Valve, and Foot of Plumber's Force Pump

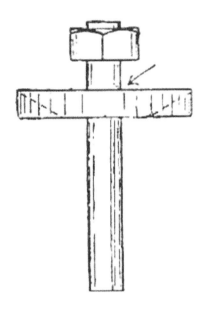

Figure 112 Jumper for Valve

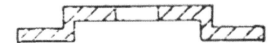

Figure 113 Section and Plan of Guide for Valve

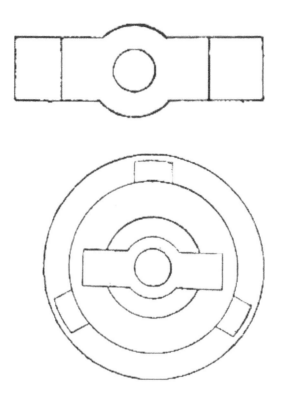

Figure 114. Section and Plan of Guide for Valve

Figure 115 Plan of Foot of Plumber's Force Pump

valve as shown in Figure 111. This grinding is done with fine emery powder and oil, very carefully, to ensure a tight valve. When this is done, the short end may be cut off as indicated in Figure 111

The guide for the valve (Figures 113 and 114) is made from a piece of 1/8-in. brass, 3/8 in. wide, bent as shown, and soldered on to the 1-in. to 3/4-in. bush-piece. Figure 115 shows the underside of the valve.

The foot to protect the valve, and the stand for the pump, are made from brass 1/8- in. thick, cut out as shown, and bent cold at the arrow-heads (see Figure 116) to fit on to the shoulder of the 1-in. to 3/4-in. bush- piece. The valve is put through the hole (see Figure 111) and screwed into the 1 1/2-in. to 1-in. bush-piece, the legs being then bent and soldered on to a 2 1/4-in. brass sink grating, or circular disc of brass. The

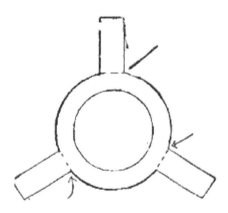

Figure 116 Plan of Foot of Plumber's Force Pump

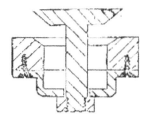

Figure 117 Section of Lead and Iron Union

Figure 118 Section of Delivery Valve

Figure 119 Section and Plan of Guide for Delivery Valve

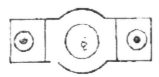

Figure 120 Section and Plan of Guide for Delivery Valve

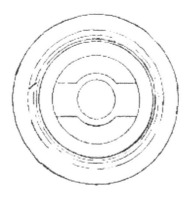

Figure 121 Plan of Lead and Iron Union

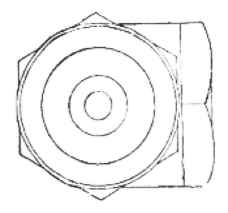

Figure 122 Lead Pipe

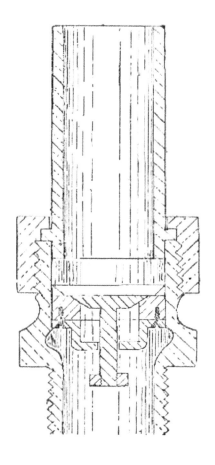

Figure 123 Section of Pump on Line A B (Figure 111)

section (Figure 111) clearly shows the mode of fixing.

The delivery valve is made up as follows: —Figure 117 shows a 1-in. lead-to-iron union, and the valve made up and soldered in. Figure 118 shows the valve open and in section. A circular piece of brass, 3/8 in. thick, and with a bore of nearly 3/4 in., and tapered inwards on the top side for the seating of the valve, is tinned on the top outside edge for soldering to the body of union. The jumper is 3/4 in., out of a 3/4-in. screw-down bib-cock, and is cut on the edge for grinding in as before mentioned.

Figures 119 and 120 show in section and plan the valve guide, which is made of 1/8-in. brass, with a hole countersunk at each end to fix into the under side of the seating; it is bent as shown, and fixed with two screws. After being ground in, this valve should be soldered in as shown in section (Figure 117).

Figure 121 shows a plan of the underside of the lead-and-iron union, with the valve and guide. In order to fix the delivery valve on to the side in the position shown in Figure 107 a 1-in. nipple and a 1-in. brass elbow with a 1-in. gas-thread are screwed on as shown in Figure 107. A 2-ft. piece of three-ply rubber pipe is fastened on to the union with fine copper wire, and a piece of 3/4-in. lead pipe tapered to 1/2-in. bore, as shown in Figure 122, and fastened with copper wire as in Figure 107.

In putting together the force-pump, all joints must be made tight, a sufficient amount of grease to ensure easy working of the suction rod being applied. A 1/16-in. hole is made at the top of the tube (Figure 107) to afford a vent for the air on the up-stroke of the suction rod. Figure 123 shows a section on the line A B (Figure 111).

CHAPTER VIII

WOODEN PUMP

The following are suggestions as to how a wooden pump may be made. Obtain the butt-end of a small elm tree, about 5 ft. long and 12 in. in diameter, and bore a straight hole 4 in. in diameter through the centre of the tree to within about 1 ft. of the bottom; the remainder of the hole should be 3 in. in diameter. With a tapering auger, take off the inside shoulder of the 3-in. bore to receive the sucker valve, as shown in Figure 124. The outside of the tree, which may now be called the barrel of the pump, can be rounded, squared, or left in the rough as may be desired. The bottom end of the barrel must be tapered outside to socket into the suction tube. The suction tubes, which should also be made out of elm, should have 3-in. holes bored through them, the ends prepared for socketing into each other, and the mouthed ends fitted outside with iron-hoops to prevent splitting. Bore a hole in the barrel, and fix a wooden nozzle in the position shown in the illustration.

The barrel is to be fixed by nailing or clamping it to pieces of wood resting upon the ground; strutting pieces must be added to hold the pump steady. Cross pieces should be attached to the suction, the ends of the cross pieces resting upon or being built into the well steining. The pump handle should be made of oak or ash. The bucket and sucker should be made of elm, but these and the bucket rod had better be bought of a pump maker, or of dealers in plumbers' materials, with the leathers and clacks fitted ready for fixing.

The sucker must be wound with hemp saturated with tallow, and wedged, by driving down, in the bottom of the barrel. The joints of the suction tubes should be put together when in a dry condition, as the swelling of the wood by wetting will afterwards make them sound. If the joints have to be remade at a future time, the spicket ends should be wound round with hemp, and then saturated with melted tallow, and inserted in the socket end before the tallow congeals. All the dimensions can be taken from the illustration.

Another method of making a wooden pump is as follows: —A tree about 25 ft. long is required; a line is struck down the centre of one side, and another line on the adjacent side at right angles. The tree is fixed in a horizontal position, about 3 ft. from the ground. At the end where the boring is to commence, two planks are laid down about 3 ft. apart, the tops being parallel to the centre line marked on the side

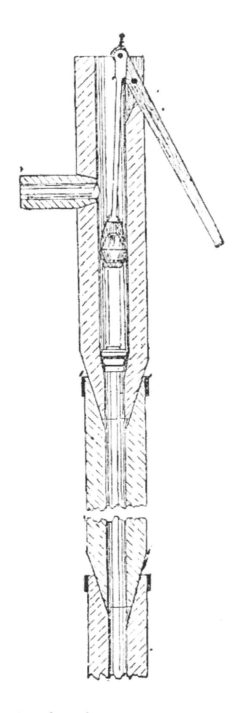

Figure 124 Section of Wooden Pump

of the tree. On these are fixed two trestles with wooden bearings on the top, in which the shank of the boring tool revolves. The larger hole for the working barrel is first bored, then finished with a smaller hole for the suction pipe. The boring tool has a thick

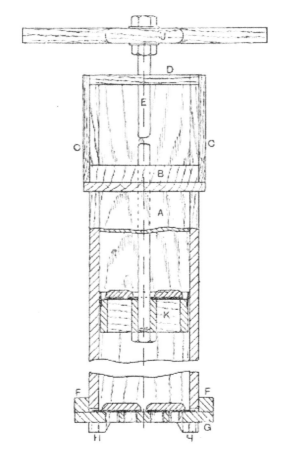

Figure 125 Front Elevation of Square Wooden Pump

shank to prevent it sagging, and is tested for being in line from the centre lines marked on the two sides of the tree. The lower end is generally set on a stone, and holes are bored round the sides to admit the water. A shell bit is used.

Pumps are sometimes made square, from 1-in. boards bolted together outside, a smaller box being fixed in the lower end to form the suction pipe. The joints must, of course, be made perfectly tight.

A pump that can be rapidly and cheaply con-

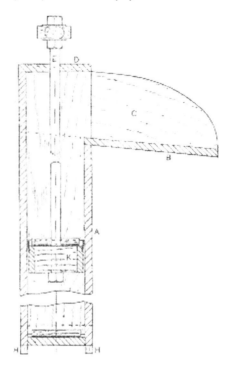

Figure 126 Vertical Section of Square Wooden Pump

structed is often required for such a purpose as emptying a flooded cellar, where a fairly large quantity of water is to be raised a few feet. For such a pump the barrel may be formed by nailing together, to form a square tube, four boards planed inside, one of these boards a (Figures 125 and 126) being shorter than the other three. Fitted to the top of a is the sloping piece B (Figures 125, 126, and 127), which, with the wing pieces C, forms the spout. A piece D is fastened across the top and serves as a guide to the pump rod E.

The suction valve is attached to the bottom end of the barrel, and is formed in the following way.

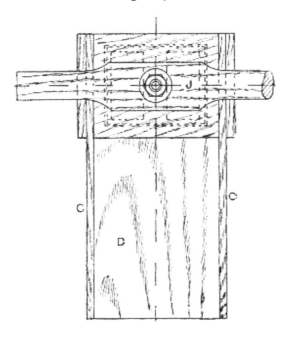

Figure 127 Plan of Square Wooden Pump

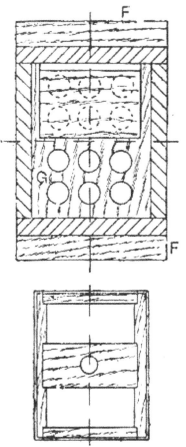

Figure 128 Horizontal Section below Bucket of Pump

Figure 129 Plan of Bucket with Valves Removed

Two battens F (Figures 125 and 128) are screwed on opposite sides of the barrel, and attached to these in such a way as to close the lower end is the piece G, in which are bored a number of centre-bit holes (see Figure 128), which are covered by the suction valve or valves.

Two valves are shown in Figure 125, each being made of thin flexible leather, or of the indiarubber sheet used in making steam-pipe joints. For a rough, temporary job, even a piece of old cloth would answer the purpose. Each valve has one edge nailed down to the plate G, and each is stiffened by wooden plates. To prevent these plates warping when wet, each must be stiffened by making it in two thicknesses with the grain crossing at right angles. Short horns H (Figures 125 and 126) are made on the ends of two of the pieces forming the barrel. These horns prevent the plate G resting on a flat surface in such a way as to choke the suction.

To the pump rod E is fastened at the top a cross- bar handle J (Figures 125 and 127) by means of nuts and washers, and to the lower end is fastened the bucket K.

The bucket (shown in plan in Figure 129) is a wooden frame which passes easily along the barrel. The top edges are rebated, and in the rebates are fastened strips of leather to prevent leakage.

On top of the bucket are two valves, which are constructed in a similar manner to the suction valves. The hinges of the valves are formed on the central cross-piece of the bucket, to which is also fastened the pump rod. This rod may be made of gaspipe, or 1/2-in. or 5/8-in. round rod, and its length should be such that when the upward stroke is completed the top edge of the bucket is not less than about 6 in. below the level of the spout; from 3 ft. to 4 ft. would be ample.

The work required to operate the pump depends on the cross section of the barrel and the height to which the water is lifted.

Before use the barrel should be filled with water. For temporary work it is not necessary to use very great care in construction, since a slight leakage will not greatly impair the efficiency. It is evident that the pump must be fixed in some way before it can be used, but the best method is necessarily dependent on the circumstances of each particular case.

CHAPTER IX

SMALL PUMPS FOR SPECIAL PURPOSES

The accompanying drawings (Figures 130 to 133) show a small pump that will be suitable for the fountain of a rockery or aquarium. It is simple in construction. The impeller is 2 in. in diameter, and the water enters on one side only. On constructing the impeller no difficulty will be experienced in making the other parts, the drawings being to scale. Figure 130 is a sectional elevation of the pump; A B is the outer case

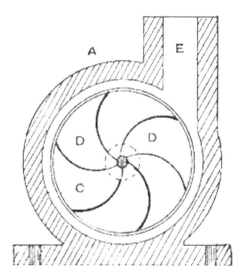

Figure 130 Sectional Elevation of Aquarium Pump

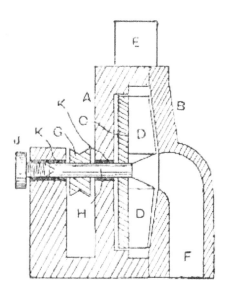

Figure 131 Aquarium Pump with Outer Case removed

constructed of white metal; C is the impeller, which is made of a disc of brass with the fans D soldered to the disc; E is the outlet or discharge; F the water inlet; G the driving pulley; and H the shaft. It would be advisable to bush the bearings with brass tube shown at K. The back bearing is provided with a set- screw J to act as a thrust bearing; the water entering the inlet will have a tendency to push the impeller backwards on the case A. Figure 131 is the pump with the outer case removed, showing the angle of the fans. The letter references in Figure 131 correspond with the same parts in Figure 130. Should there be any difficulty in making the entire case of white metal, it may be constructed of wood in three parts screwed together as shown in Figure 132. Figure 133 shows the back half a of the case. The back half is hollowed out to take the impeller, allowing about 1/8-in. clearance all round the circumference of the impeller as shown. The front half B of the case is also hollowed out. The space between the sides of the case and the impeller should be as small as possible— just sufficient for the impeller to revolve without touching. The

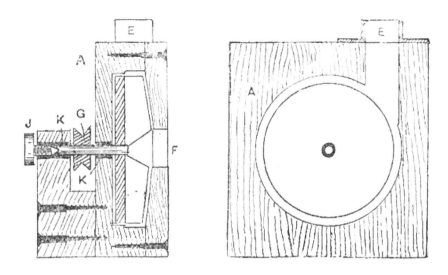

Figure 132 Aquarium Pump with Wooden Case

Figure 133 Back Half of Wooden Case

inlet pipe is not shown in Figure 132, only a hole through the case. A pump of this size will not draw its own water satisfactorily, therefore it is necessary to fix the pump below the water level in the basin, so that the water may run in by gravitation. The number of revolutions to raise the water 3 ft. high will be approximately 2,300 per minute. The power required to drive this pump is but small, the load being only about 240 foot-lb. —say 300 foot-lb.; this is equivalent to about 10 watts. This is the actual electric power required to be developed by the motor, and does not take into account the efficiency of the motor.

A pump for testing purposes, for feeding a small boiler, and for many other purposes, may be of the type shown by Figure 134. This is designed to fulfil all the requirements of the orthodox hydraulic testing pump of very costly design and workmanship, and can be built for a shilling or so. It consists of a rod of solid brass or gun-metal drilled out to the required sizes as shown in Figure 134, and fitted with ordinary iron gas or steam bends at the top and

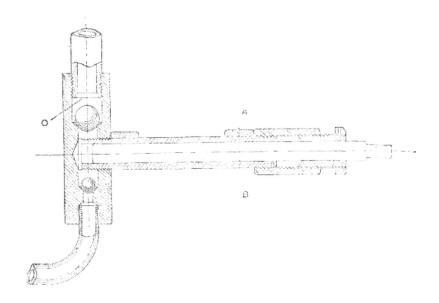

Figure 134 Section of Hydraulic Testing Pump

Figure 135 Clamp for Hydraulic Testing Pump

Figure 136 Upper Pipe with Cross-bar

bottom of valve box. The upper pipe carries a small cross-bar to keep the discharge valve from blocking the outlet C (Figure 135). The two valves are the ordinary playing marbles (glass preferred, as these are far more perfect in shape than the clay ones), which drop into the seatings made by the drill. The centre of the piece of gun-metal or brass is drilled and tapped to fit a short length of steam or gas barrel, the length of which is determined by the length of stroke the pump is to have. The gland end of this piece of tubing is countersunk as shown, and an ordinary socket fits over this, wherein is screwed the gland, which consists of an ordinary nipple, also countersunk; and for screwing up the gland a back nut is fitted, and secured by a rivet to keep it from turning. Before fitting the pump plunger it is as well to file the barrel out roughly, to get rid of any scale or

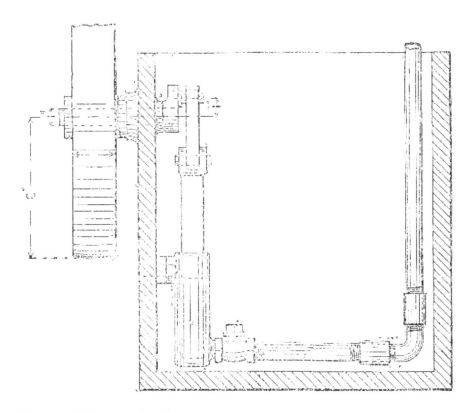

Figure 137 Lubricating Pump

inequalities left from the pipe mandrel, and the plunger can be turned to fit this loosely; one end of the plunger has two flats filed on it and a hole drilled for a pin, and the pump is complete. For securing in position, two ordinary bent clips embrace the barrel near the ends as shown in Figure 136, and in section through A B (Figure 134). This pump would stand a pressure of 250 lb. per square inch easily, and would do the work of one costing ten times as much; while

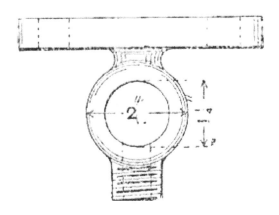

Figure 138 Barrel for Lubricating Pump

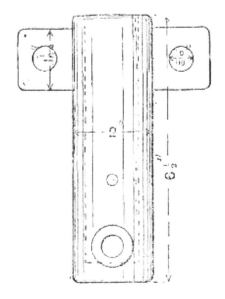

Figure 139 Barrel for Lubricating Pump

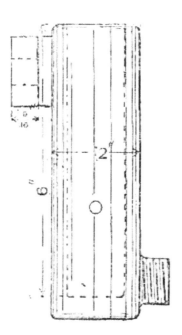

Figure 140 Barrel for Lubricating Pump

valve used being an ordinary 1/2-in. brass check or nonreturn valve, which is screwed to the pump nozzle, as shown at the bottom of Figure 137, and coupled to the pipe through which the oil is forced. The pump barrel (Figures 138, 139, and 140) is bored 1 1/4 in. by about 6 in. long, and has an outlet at the bottom at right angles to the bore; this outlet is threaded to suit an ordinary 1/2-in. gas coupling, and should therefore screw into a standard valve. The ram (Figures 141 and. 142) is turned a good fit in the barrel, and has a stroke of 2 in.; therefore the crank-pin centre will be 1 in. from the centre of the driving shaft, this having a pulley keyed on its outer end. The small crank can be forged solid on the shaft, and the crank pin should be 1/2 in. in diameter and screwed into the crank. A small hole 1/8 in. in diameter should be drilled near the end for a split pin which will prevent the connecting rod coming off when working. The connecting rod consists simply of a piece of wrought- iron I 1/4 in. wide by 1/2 in. thick, with a hole drilled at each end, one for the crank pin, the other to be coupled to the cross-head. The pump should be fas-

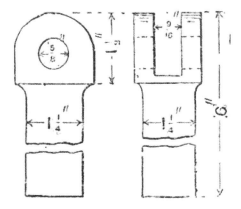

Figure 141 Plunger for Lubricating Pump

Figure 142 Plunger for Lubricating Pump

tened by 1/2-in. set screws inside a cast-iron tank, say 3/8 in. thick and about 16 in. long by 9 in. wide and 18 in. deep. The tapped holes in the tank must be made oil-tight. The tank should have sufficient oil poured in to cover the pump body, and, in fact, the oil should never be allowed to get below the top of the pump body. When at the top of its stroke, the bottom of the plunger should be just above the 1/4-in. holes (four in number) drilled through the pump body, and shown in Figures 138 and 139. The oil will rush through these holes into the barrel, and, on the downward stroke of the ram, be forced through the outlet at the bottom and up the supply pipes. It will be prevented from returning by the non-return check valve. The speed of the driving pulley should be about twenty-five revolutions per minute. The pipes are ordinary 1/2-in. gas-tubing, but could be reduced to 1/4-in. piping at the bearing to be lubricated, and an ordinary gas tap may be screwed, on so that the oil can be turned on or off at will. In this case an outlet pipe should be placed in the service pipes to keep them from bursting when the pump is working. This outlet or relief pipe should convey the oil back to the tank to be used again, which is also the case with all the oil used, for when it leaves the bearing it should be collected in oil cups cast on the bottom of the bearing pedestal, whence it is returned through a 3/4-in. gas pipe back to the oil tank. These pumps can be arranged in series or gangs of any number inside one tank, and can all be driven by one shaft, the pumps being placed on one side of the tank at equal distances apart. The first pump is driven by a pulley in the ordinary way, but, instead of a crank to give motion to the ram, it has a small spur wheel with the driving or crank pin screwed into it 1 in. from the centre; this spur wheel gears with another one of equal size which drives the next pump, and so forth to any number.

CHAPTER X

CENTRIFUGAL PUMPS

Dimensioned drawings of a centrifugal pump designed to lift 150 gal. per minute at 20-ft. head are presented by Figures 143 to 146. To enable the volute to be correctly formed the case is in two halves.

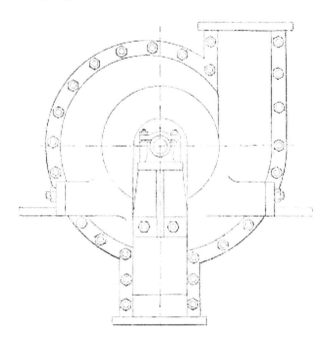

Figure 143 Side Elevation of Centrifugal Pump

To avoid end thrust and to ensure an even balance of the disc, the inflow takes place on each side, each inlet having a diameter of 3 in. Figure 143 is a side elevation and Figure 144 one half of the case showing the depth of the volute and disc with angle of vanes. The volute, to obtain a good flow, must increase evenly

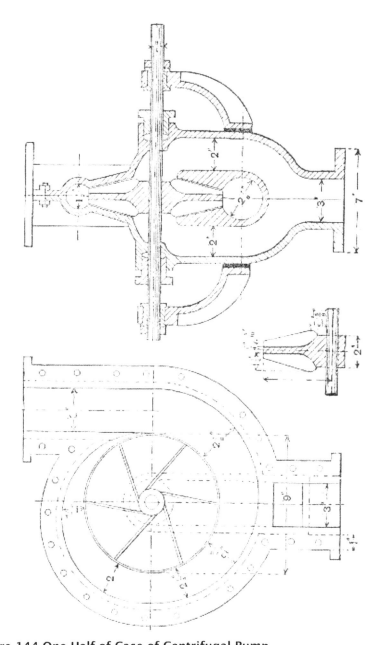

Figure 144 One Half of Case of Centrifugal Pump

Figure 145 Cross Section of Centrifugal Pump

Figure 146 Part Section of Disc.

to its discharge. The discharge pipe should increase in area to reduce the velocity considerably. The flange of the casing is 1 in. wide, drilled to take −3/8 in. bolts. The diameter of the disc is 9 in., and is arranged for six vanes, having an angle of 80° at the cir-

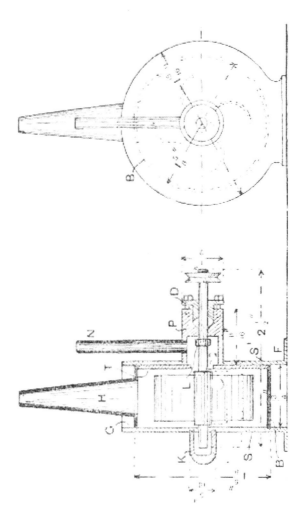

Figure 147 Vertical Section of Model Centrifugal Pump

Figure 148 Side Elevation of Model Centrifugal Pump

cumference. The shaft is 7/8 in. diameter, and the approximate speed of disc is 650 revolutions per minute. Figure 145 is a section showing side inlets, disc, and brackets, and Figure 146 is a section of half of the disc showing dimensions of the vanes.

The model centrifugal pump illustrated in section by Figure 147 and in side elevation by Figure 148, was made with materials that happened to be at hand, no castings being used. Brass plate, a piece of brass tube about 1 5/8 in. in diameter and 3/4 in. long, and some brass rod were the only requirements.

The tube should be chucked in the lathe and the edges turned true and square to form the barrel B (Figures 147 and 148). On brass plate about 1/16 in. thick, two circles 2 in. in diameter should be scribed and cut with a fretsaw to form the sides shown in Figure 148, the edges being finished with a file. Leave sufficient metal to bend at right angles to form the feet F (Figure 147). A 1/8-in. hole being drilled exactly in the centre, these sides can be chucked in the lathe by being screwed to a disc of wood.

In the side s turn a slight groove G to fit the barrel; this ensures the shaft and vanes being central with the barrel, which is important, besides making a better soldered joint.

A brass, disc T 1 5/8 in. full in diameter should be cut, centred carefully, either screwed or soldered to the side S', and turned to fit tight inside the barrel so that with a little white-lead the joint will be quite firm and watertight without being soldered. The hole in the outside plate must be enlarged to about in. for the stuffing-box p.

The stuffing-box is made from brass rod 1/2 in. in diameter and 3/4 in. long, chucked in the lathe, a hole 1/8 in. in diameter being bored (preferably with a twist drill) right through, this hole being enlarged to about 5/16 in. for a length of about 1/4 in., for the gland D. The last is made from brass rod, and has two small holes for screws, corresponding holes being drilled and tapped in the stuffing-box. The 1/8-in. hole must be enlarged at the other end of the stuffing-box to about 3/8 in. by 3/8 in. long, and a hole drilled to take the 1/8-in. inlet pipe N, which is a piece of 1/8-in. brass tubing, screwed or soldered in place. The stuffing-box is turned on the outside and fits tight in the hole in the side, but is not soldered until all the parts are finished. In the feet, two holes are drilled for screws for fastening to the baseboard.

The shaft is made from a piece of 1/4-in. brass rod, 2 1/8 in. long; it should be turned down to 1/8 in. for lengths of 1 1/4 in. at one end and 5/16 in. at the other, so as to fit the hole in the stuffing-box without shake. A collar L must also be turned on, or a loose one soldered in place, the shaft being allowed a very slight side play in the bearings. Four equidistant slits parallel to each other should be cut in the boss on the shaft for the vanes. These are cut from hard brass about 1/32 in. thick, bent to shape (Figure 149) and then soldered in place.

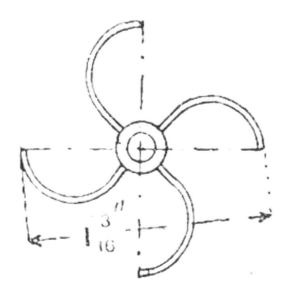

Figure 149 Vanes for Model Centrifugal Pump

A small pulley about 3/8 in. in diameter is turned and grooved and screwed or soldered on the shaft at the stuffing-box end. To prevent leakage of water at the other end, the cap K must be turned and bored, rounded, and soldered in place exactly over the hole for the shaft. The 1/8-in. hole in the side plate T (Figure 147) is enlarged a little to allow the water to flow in. A hole not less than 3/8 in. should be made in the top of the pump barrel for the discharge pipe H.

The discharge pipe is made from thin brass or copper, and soldered in place, care being taken that it does not project inwards so as to touch the vanes when revolving.

The parts may next be fitted together, the barrel being first soldered to the side S; for this the best method is to wet the joint all round with killed spirits of salt, putting a few specks of solder on and holding over a Bunsen burner or spirit lamp, when the solder will flow round the joint, all superfluous solder being removed. The stuffing-box is also soldered in a similar manner to the side S', care being taken to fix it perfectly central. The shaft may then be put in place, the stuffing-box packed with a little cottonwool and oil, and the gland screwed up. The other side plate, after being smeared with a little white- lead, is snapped in place, and if accurately fitted will be quite firm and watertight and can be removed at any time with a few light taps of a hammer.

An indiarubber tube should be attached to the inlet pipe and connected to a vessel of water; also, if the pump is intended to work a model fountain (which it is most suitable for), a conical tube ending in a fine jet must be attached to the discharge pipe by means of an indiarubber tube. When connected to a lathe wheel this little pump would throw a jet of water fully 6 ft. high, and when driven by a small electro-motor would throw a jet 2 ft. high, which would be sufficient for a model fountain.

CHAPTER XI

AIRLIFT, MAMMOTH, AND PULSOMETER

The air-lift pump is very valuable for pumping from bore-holes, and is not limited to this kind of well, being equally useful for a dug well, sump, etc., the

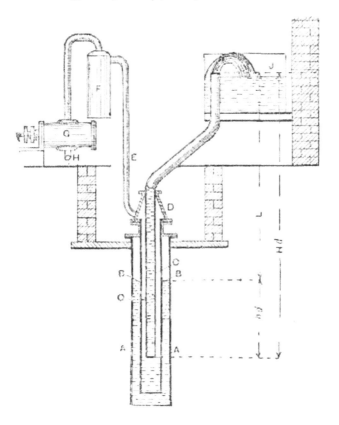

Figure 150 Typical Arrangement of Air-lift Pump

only condition being sufficient depth of water to submerge the air and delivery pipes for about half their total length, or rather more, according to the lift. The general arrangement of an air-lift pump is shown by Figure 150, in which A is the outer tube of an artesian well, down the centre of the well being another tube B; this tube must be of sufficient diameter to leave an annular space round it and the outer tube as shown, terminating in a cap D at the surface. Through this cap passes the pipe C, which is continued down the outer tube of the pump b to nearly the bottom. From the cap D rises the pipe E, which connects the cap with an air-receiver F. G is the cylinder of the air compressor, H being the suction pipe. The central tube C is continued to the tank J, and acts as the rising main. From the tank the water is conveyed to the storage reservoir, this type of pump not being suitable for discharging into a length of pipe.

The action of the air-lift pump shown by Figure 150 is as follows: Air is pumped into the receiver e by the compressor G, which may be actuated by steam or any other motive power. From the receiver the air is conveyed to the annular space between the inner pipe C, which forms the rising main for the water, and the outer pipe B, by the air pipe E. The effect of this pressure is to make the water rise up the central pipe C, and this continues until the water-level in the outer pipe B has descended to the level of the bottom of the tube C. Air then escapes up the tube C, taking the water with it, and lifting it to the desired height. The size of the central pipe C, and the depth to which it must be taken down, are points which have to be carefully arranged to suit each particular case. It must be of the right diameter, having regard to the quantity of water to be raised, and the amount of submergence found necessary determines the air pressure required.

To illustrate the theoretical and practical limitations of the above system, hd (Figure 150) represents the depth from the water surface to the point of application of the air, H d the head of water above the point of application of the air, and L the lift against which the water is pumped. In Figure 150, which represents the normal depth of submergence to obtain the highest efficiency, hd is about 0.6 of Hd; theoretically it should be 0-5 of Hd, the reason for this depth of 0.6 of Hd being that the pressure at the point of application of the air due to the head hd should be slightly in excess of the pressure in the discharge pipe. The necessary air pressure required to raise the water is equal to the depth of submergence of the pipes. As soon as the pressure of air in the supply pipe has been raised to that of the water at the foot of the well, the pump begins to give off water. Take, for instance, a case where the lift is 80 ft., and the corresponding immersion is 120 ft., thus making the total length of the rising main 200 ft. To overcome the 120-ft. resistance of immersion, the air must be compressed to 60-lb. gauge pressure or 5 atmospheres absolute pressure. This pressure of 60 lb. is exactly equivalent to the water-level in the well.

Another arrangement of air-lift pump is shown by Figure 151. a represents the depth from the point at which the water is to be delivered to the point of application of the air. B represents the head of water when pumping above the point of application of the air; and C represents the lift against which the water is pumped. The diagram shows about normal conditions: B is about 0.6 of A; theoretically B should be 0.5 of A, because the pressure at the point of application of the air due to the head B should be equal to or slightly in excess of the pressure in the discharge pipe at that point. The air pressure in pounds per square inch required is therefore equivalent to the pressure of a column of water, the length of which is equal to the depth to which the air pipe is submerged.

A modification of the Shone pneumatic system of raising sewage is illustrated by Figure 152. This may be employed for raising water from any depth. In Figure 152 a is a metal cylinder with a valve B opening inwards; C is the rising main, which passes through the top of the cylinder reaching to within about 1 in. of the bottom; D is the pipe for the compressed air; E is a three-way cock, one way opening to the atmosphere, and the other to the compressed air supply. The water is raised by turning the tap E so that the atmosphere is in communication with the cylinder, when the water rushes through the valve

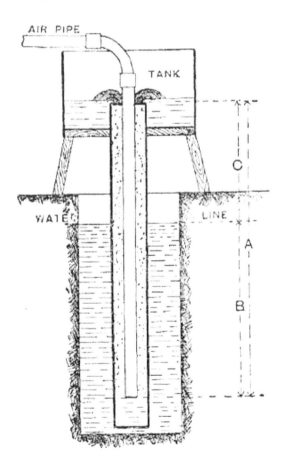

Figure 151 Another Arrangement of Air-lift Pump

filling the cylinder. The tap is now turned to put the compressed air in communication with the cylinder, and the pressure on the surface of the water forces it to the surface. To continue the action the cycle is repeated. The tap may be worked automatically by means of a float in the cylinder.

A type of pneumatic-acting pump used in some parts of Hungary is shown by Fig; 153. This is speci-

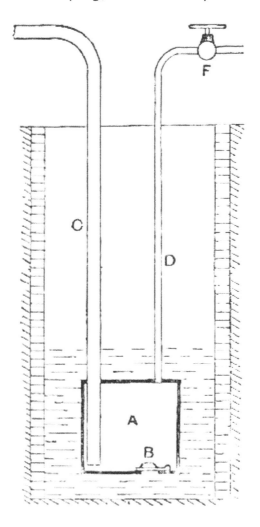

Figure 152 Modified Shone Pneumatic Pump.

ally adapted for deep borings. It is connected with a barrel or chamber a sunk into the liquid to be raised, in such a manner that this chamber may, by turning the three-way cock B or any other reversing device, be put into communication with an air-pressure con-

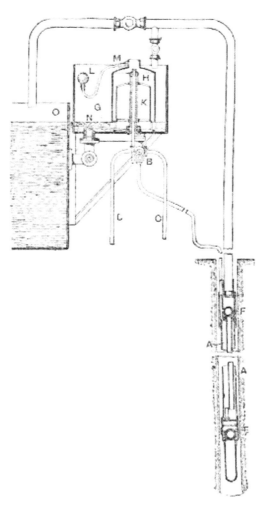

Figure 153 Pneumatic Pump used in Hungary

duit C or with the suction pipe D. The barrel or chamber a, which may be constructed of some flexible material if for bent borings, is fitted with a suction valve E and a pressure valve F, and when it has been immersed in the liquid to be raised and put into communication with the suction pipe, it will become filled with liquid. Part of the liquid raised is discharged into a tank G, containing a bell H in which is a float K. On the tank becoming nearly full, the float L opens the air-escape valve M, allowing the liquid to enter the bell and raise the float K. The three-way cock B is by this means turned so as to put the forcing chamber in communication with the air suction, and the valve N is opened to allow the liquid in the tank G to flow into the receiving tank O. When the level falls below the mouth of the bell, the float K drops, closing the valve M and again putting the forcing chamber in communication with the air pressure. The mouth of the bell is preferably on a level with the highest level which is to be attained in the tank O; the action of the apparatus is thus automatically arrested when this level is attained. A float valve may be provided in the bottom of the chamber to prevent loss of air if the chamber is emptied before the air supply is cut off.

In a modification of the above system the forcing chamber is comprised between two pistons movable in the tube sections of the boring. One of these pistons is connected to an air pipe and separated by another pipe from the second piston, which is supported by a wire rope. On this rope being tightened, the rubber packing rings on the piston are tightened against the boring casing. Air is supplied to the air space between the pistons by means of pipes and the annular space between the outer and inner packing rings of one of the pistons, and water is discharged by the pipe which separates the pistons, which communicates with the space surrounding the air pipe.

In another modification suitable for wells of comparatively large diameters, a system of floats and valves similar to that illustrated is arranged within the forcing chamber, which is relatively short and of correspondingly enlarged diameter. Instead of the air supply to the forcing chamber being controlled by the discharge liquid, it may be controlled through gearing or the like from the air compressor.

The apparatus known as the Mammoth pump is not yet much in use in the United Kingdom, but there appears to be as large a field for it here as elsewhere. It is much more simple in construction than an ordinary pumping apparatus, takes up less space, costs less money, and up to certain limits has greater practical advantages than the common pump, and, indeed, will in some cases work where ordinary pumps fail.

Although it was invented more than a hundred years ago, it is only within quite recent years that it has been developed and perfected. A glance at Figures 154, 155, and 156 will explain the working of the apparatus. A indicates the tank or well from which the fluid is to be drawn, b the rising main, C the air-pipe, which is carried round into the foot of B so as to ensure that the air escaping from the orifice D shall pass up the pipe B; E indicates the imaginary standing level of the fluid to be pumped; F the level of the invert of the discharge of the pipe B; and G the tank into which discharge is to be made. There are, of course, various modifications possible, but the diagrams have been kept simple in order to explain the principles which govern the successful working of the apparatus.

Under normal conditions the fluid, as already stated, stands at level E; but as soon as air is forced through the pipe C the equilibrium of the fluid is disturbed, that outside the tube b remaining at its original weight, whilst that inside the tube (being filled with small bubbles of air) has its weight much reduced.

The result is that the pressure of the external fluid forces the lighter internal combination of fluid and air up the rising main until it flows over the invert at level F into G, which may be a tank, or a race, or similar contrivance to separate the air from the fluid.

The orifice D may be a plain open end to the pipe C, but better results are obtained by using a perforated ring, or cross, or a combination of both, the object being to divide the air up into as many small bubbles as possible, so as to ensure its more equal distribution, and therefore a more perfect mixture of air and fluid. When this is done it is advisable to put an enlarged end to the pipe so as to avoid obstructing the flow of the fluid into the bottom of the rising main. Figure 155 shows one way of carrying out these ideas.

It will be seen on consideration that the question of depths from which, and heights to which, fluids may be raised depends on the density of the liquid, the quantity and degree of compression of the admitted air, and the depth of submersion of the rising main. This latter is the most important item, as the depth of submersion forms the head of pressure which is required to force up the combination of air and fluid. With ordinary water it may be taken as equal to one and a quarter times the height of lift (that is the difference between the levels of E and F); but, of course, a submersion of one and a half times the lift would give a more rapid flow, and some engineers prefer to work to these proportions; but the mixture of air and water will actually rise in considerable volume when the subversion is a very little more than the height through which the fluid has to be lifted.

The quantity of air required to be admitted varies according to its degree of compression. It is usually, for fluids with the specific gravity of water, simply forced in at the atmospheric pressure, and then the

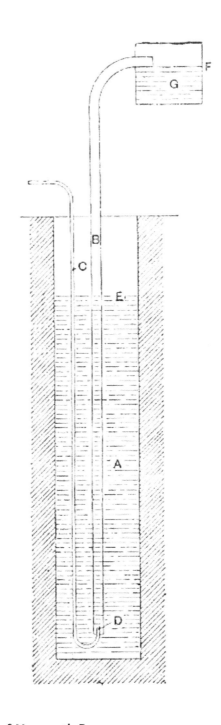

Figure 155 Section of Mammoth Pump

necessary quantity is about three times the bulk of the fluid raised, the depth of immersion being one and a quarter to one and a half times the height of lift. For fluids denser than water it is necessary to compress the admitted air (and, indeed, better results

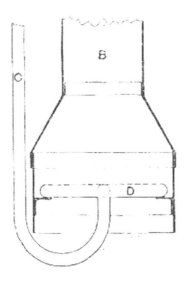

Figure 154 Section through Enlarged End of Mammoth Pump to Rising Main

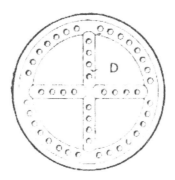

Figure 156 Plan of Enlarged End of Mammoth Pump, showing Perforated Orifice to Air-pipe

are obtained in lifting water when the air is supplied under pressure), the pressure varying from two to twelve atmospheres.

In the case of hot fluids the efficiency of the mammoth pump is much increased, as the admitted air expands considerably in volume as its temperature is raised. As is well known, the exact opposite is the case with ordinary suction pumps, which are deteriorated by hot fluids; and, in some cases, altogether fail to perform the work required of them.

The great advantages of this form of apparatus are also apparent for lifting water charged with salt or other chemicals, as there are no working parts to become clogged or deteriorated. There is nothing inserted in the fluid, or in any position which the fluid can reach, except two pipes and an air nozzle, all of which can be replaced rapidly and cheaply, so that a great saving of time and money may be effected in this respect.

It is also possible to drive the pump from machinery situated at a distance, and to start the lifting of the fluid by merely turning a tap. This is a very great advantage, as the air pump may be placed in the shop and driven off shafting in connection with other machinery, thus avoiding separate plant for raising steam, or a long and costly series of steam pipes with the accompanying loss of heat. There are also other advantages which are so obvious as to need no mention. The one great drawback is that a tank cannot be pumped perfectly empty by its means.

A Pulsometer pump (manufactured and sold by the Pulsometer Engineering Co.) has nothing in common with the other devices described in this chapter beyond the object for which it is used. It is an appliance for raising water by the alternate pressure and condensation of steam. To describe the parts shown in Figure 157, J is a pipe from a boiler containing steam under pressure. The gunmetal spherical valve is free to move and alternately to cover the necks C and B.

The latter form the upper parts of the chambers A A, into which water passes through the valves E E from the suction pipe D. H H are doors for access to the valves E E for repairs or other attention. Near the bottom ends of a a are side outlets, as shown by the dotted circles, covered by the valves F F, also shown

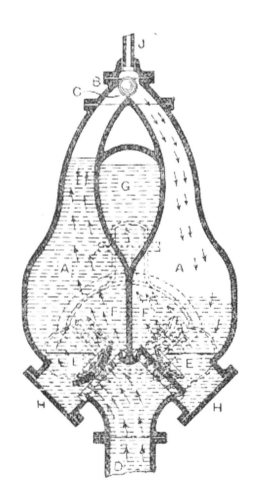

Figure 157 Sectional Elevation of Pulsometer

by dotted lines, opening into a chamber with which are connected the air vessel B and the outlet branch to which the delivery pipe is attached.

The action of the Pulsometer is as follows: —The pump is first charged with water through plug-holes provided for the purpose, and then steam is turned on at j

J. This presses on the water on the right hand chamber a (which is not covered by the spherical valve), and forces it, as shown by the arrows, through the right-hand valve F and up the delivery pipe. The steam in the right-hand chamber a then condenses, and causes the spherical valve to roll over and cover the opening to this chamber, and also creates a vacuum, which is again filled with water through the right-hand valve E from the suction pipe D. When the valve has rolled over D, the steam passes through the open neck C, and presses on the water in the left-hand chamber A, forcing it through the dotted left-hand valve F into the delivery chamber. When the left-hand chamber a is nearly empty, the valve is again pulled back by the condensation of the steam in the chamber, which again fills with water during the time the other chamber is being emptied, and these actions continue so long as steam under efficient pressure is supplied.

As water will not rise in a vacuum beyond a certain height, a pulsometer should not be fixed more than about 15 ft. or 20 ft. above the water to be raised, although theoretically the limit is a little more than 30 ft.

The Pulsometer pump can be slung on chains in a well or sump, so that there is very little trouble in fixing it, or lowering it when necessary for keeping within the above distance above the water. The height to which a Pulsometer will raise water depends on the pressure of steam in the boiler. Its special features are its portability and its capacity for pumping dirty liquids.

CHAPTER XII

HYDRAULIC RAMS

Hydraulic rams are water-raising appliances in a class by themselves. The shock that is commonly noticed on quickly closing a fullway cock and suddenly stopping the flow of water in long lengths of pipe is the power employed in the hydraulic ram.

The hydraulic ram was invented in 1772 by Whitehurst, and about the same time it was accidentally discovered by a Bristol plumber, who was engaged in a hospital in that city fixing long lengths of lead piping, which had a considerable head of water on them. To the extreme end of one of the pipes he was fixing, and at the lowest point, he had soldered an ordinary ground bib tap, with a plug that was very easily turned. When turning off the tap quickly, he found that the pipe had split and burst. After repairing it, he found the same thing occurring again every time he closed the tap suddenly. This caused him to consider, and he came to the conclusion that the evil was caused by the sudden closing of the tap arresting the flow, or momentum, of the water in the pipe plus the weight of water in the length of pipe, exerting a blow, generated by the excess of pressure and causing the pipe to burst. After trying many experiments in order to remedy this defect, the plumber soldered a small pipe behind the tap, carried it up the wall, and discharged it over the top of the cistern that supplied the tap. Every time the tap was used, he found that the water rushed back into the cistern with great force and noise. As this seemed very strange, he determined to test the matter, and he then continued the pipe up to a cistern on the roof of the house, which he found could be supplied with water from a lower cistern by simply turning the tap quickly on or off.

It would be rather difficult to decide who was the original inventor of the ram, as it has been rediscovered and invented by engineers and plumbers several times during the past century. It was improved subsequently, and made to work automatically, by Montgolfier, the celebrated inventor of the balloon, and his son, who substituted for the tap, turned by hand in Whitehurst's ram, a large ball- valve enclosed in a cage opening outwards and closing by the momentum of the water rushing through the trunk or body of the ram, and a similar but smaller valve opening inwards inserted in the air-vessel. Since then the ram has been improved by various inventors—by Keith, Fyffe, Davies, and others— but principally by the late Mr. John Blake, of Accrington, who made it the study of his life. As the ram then was, he found it little more than a toy or scientific curiosity, only suitable for raising small quantities of water. By constantly experimenting, he added improvement after improvement until he brought it to its present high standard of efficiency — capable of economically supplying villages and small towns with water.

The hydraulic ram is a machine of very simple construction, but it does not seem to be understood as it should be by engineers and plumbers who may have the erection and installing of it. It consists of a body, or trunk, and an air-vessel, in which are inserted three valves, namely: (1) The pulse, or foot valve at end of the trunk; (2) the retaining, or ascension, valve, in the air-vessel; and (3) the snift valve, in the neck of the air-vessel, immediately under the ascension valve. The office of the snift valve is to supply air to the air-vessel when the ram is working. This, though seemingly the most unimportant valve in the ram, is most necessary to the successful working of the machine, as without it the ram would soon cease working, owing to the air having been exhausted, as the air-vessel would have become waterlogged. Air escaping from the air-vessel with the water causes the severe shocks and noise in badly constructed rams that are not fitted with this simple valve.

The hydraulic ram is a machine that utilises the momentum of a stream of water having a slight fall, in such a way as to elevate a portion of that water to a greater height. This machine is self-acting; and when once set going it will go on working day and night for a long period without stopping, provided that the supply of water is sufficient and that the ram is properly constructed and properly fitted up. The present writer fitted up, more than two years ago, a ram that has worked night and day without intermission, except when, on one occasion only, it was stopped for an hour for examination of the valves and flushing of the ram.

The hydraulic ram will force water with a fall of from 18 in. to 100 ft. to an elevation of 15 ft. to 1,000 ft. from almost any distance, from a few score of yards to four or five miles, or more if necessary. For example, supposing 100 gal. of water falling 10 ft. would elevate 10 gal. to a height of say 80 ft., as 100 gal. falling 5 ft. will elevate 1 gal. to a height of 300 ft., a hydraulic ram will raise water from 300 gal. up to 500,000 gal. per day of twenty-four hours if required.

The following example will give some idea of the power exerted at the end of a long pipe when the flow is suddenly stopped: A pipe flowing full bore with a velocity of 25 ft. per second is equal to a head of 10 ft. If the pipe is 2 in. in diameter and 150 ft. long, the 2: ' x 6^-5

contents will be $$\frac{2^2 \times 62.5}{144} \times 7854 \times 150 = 204.5 \text{ lb.},$$

which multiplied by 10 = 2,045 ft. -lb. of energy.

A diagrammatic section of a hydraulic ram is presented by Figure 158. a is an air vessel, B and C ball valves, D a delivery pipe, and E the supply pipe. Above the valve B is an opening, and the water, in running down from a small fall at E, passes through this outlet until the velocity is sufficient to close B. This, of course, suddenly stops the stream, and the outlet valve C is forced open owing to the great increase of pressure in the ram. Through C the water passes into a and up the delivery pipe D. This releases the pressure and the valves B and C fall, and the operation is gone through again. In some cases an ordinary lift or a flap valve, which must be weighted to exceed slightly the static pressure of the supply stream, is placed between E and C. Obviously, a portion only of the supply water from a small fall is delivered to a greater height, and the average efficiency of the ram is probably not more than 50 per cent.

Illustrations will now be given showing the construction of a 1-in. ram in copper sheet or tube, suitable for springs or streams supplying from 90 gal. to 300 gal. per hour and raising approximately from 5 gal. to 20 gal. per hour to a height of 80 ft.; but the illustrations are so proportioned that a ram can be constructed from them to suit any fall of the drive pipe, and any quantity of water, by simply increasing the dimensions according to the varying conditions of the stream on which the ram is to work.

The quantity of water raised by a hydraulic ram varies according to the ratio of the fall to the height that the water has to be raised, while the effective capacity is materially affected by the length of the drive and delivery pipes, as well as by the relative fall and lift.

A ram will work on as low a fall as 2 ft. 6 in., and the quantity raised will be in proportion to the fall; generally, a proportion of 1 in 10 of fall to lift gives the greatest efficiency. As there are not two streams whose conditions are exactly alike, it is obvious that a ram that will work well on one stream will not work satisfactorily on another; therefore, to obtain the highest efficiency, the ram must be designed to suit the conditions of the stream on which it is to work. These conditions are ascertained by surveying the site and making other observations. The data required are: —(1) The minimum summer quantity of water in gallons supplied by the stream per minute; (2) the horizontal distance from the stream to the proposed site of the ram; (3) the vertical height obtainable from the supply source to the ram; (4) the vertical height the water has to be raised above the ram; (5) the horizontal distance from the ram

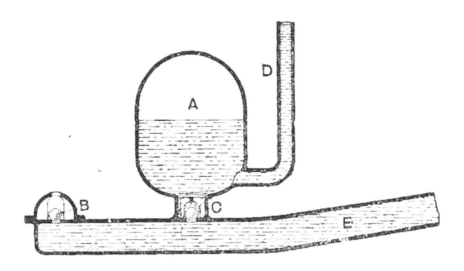

Figure 158 Hydraulic Ram: Explanatory Diagram

to the end of the delivery pipe. The method of making these observations will be described later.

The ram illustrated at Figures 159 and 160 may be constructed of copper sheet or drawn tube, the tubing being more satisfactory. To withstand the pressure, the thickness should be not less than 3/32in. for low falls of 5 ft. and a short drive pipe; but for falls above this, and long drive pipes, the thickness should be 1/8 in. Figure 159 is a section of the ram, and Figure 160 an end elevation shown with the connecting flange cut away on the line X X (Figure 159).

The body a is 5 3/8 in. long when finished, and to allow for the flanges B and C the sheet or tube must be 7 1/8 in. long; the width of the sheet will have to be sufficient to make a tube of 1-in. internal diameter, and to allow for a 3/8-in. Lap for the joint. Small rivets are used for making the joint, which is afterwards soldered. The edge of the lap on the inside should be bevelled off, and care should be taken that the heads of the rivets do not project much into the ram. The flanges are worked over as shown, to 2 3/4 in. in diameter. At 1 3/4 in. from one end a hole 1 in. in diameter is cut, the centre of the hole being 2 1/4 in. from the end; over this hole is riveted the T- piece D, communicating with the delivery valve.

The T-piece is of the same diameter as the body of the ram, and when finished is 1 in. high. The sheet or tube will have to be cut sufficiently long to allow of a flange 3/4 in. wide being worked on the end resting on the body of the ram. The T-piece is riveted to the body as shown; and to make a perfectly watertight joint, it is also soldered. To avoid having rivet heads on the inner surface of the ram, the T-piece and body are bolted together with brass wire of 1/8-in. diameter, with a fine thread cut on it—Association gauge would be suitable —the holes in the flange of the T-piece and body being tapped for this thread. When all the studs are screwed up, they are cut off nearly flush—say to about 1/64in. of the inside and outside surface of the copper—by filing; then, on giving them a few blows on the outside with a hammer, after inserting a 1-in. diameter iron bar in the tube, the ends will be burred over a little, making a sufficiently strong joint when sweated together with solder. The inside ends of the studs should then be filed off flush, to prevent obstruction to the flowing water.

The elbow E (Figure 159) is a straight piece of copper tube of 1-in. internal diameter. For rams working on even low falls the thickness must not be less than 1 1/8 in., so that the bend may be worked out to the larger diameter as shown, the elbow being enlarged where the pulse valve works to an internal diameter

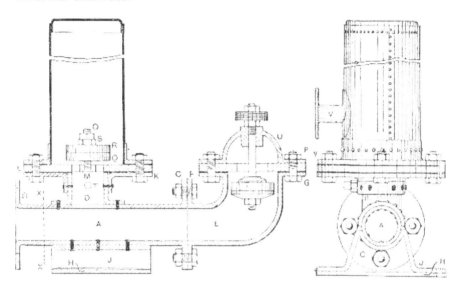

Figure 159 Section of Copper Hydraulic Ram

Figure 160 End Elevation of Copper Hydraulic Ram

of 15/8 in. The tube must be of sufficient length to allow of the flanges F and G being worked over; the flange F, which is bolted to the body, is 23/4 in. in diameter, and the flange G 3 3/8 in. in diameter. The approximate length of tube required is from 7 1/2 in. to 8 in. Holes for 1/4 in. bolts are drilled through the flanges as shown.

The bed plate H and J (see also Figure 160) is made of two pieces of sheet brass 1/8 in. thick. The bottom piece H is 31/4in. long by 4 in. wide, and to this is bolted and sweated the cradle J, made out of a sheet about 63/4 in. long by 3 1/4 in. wide, shaped as shown, to fit the body of the ram. The bed plate and body are riveted

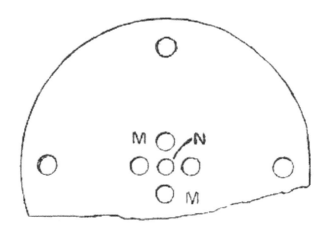

Figure 161 Plan of Delivery Valve Seat for Hydraulic Ram

together, or bolted and sweated, in the same way as the T-piece.

The flange K (Figure 159) is 1/8 in. thick by 4 in. in diameter; from the centre the metal is worked out at right angles to form a ring with an internal diameter of 1 1/4 in., this ring fitting tight over the T-piece, to which it is secured with rivets. By working out from the centre there will be sufficient width to make the rings 1/2 in. in length, which will be ample. To this flange is bolted the circular valve seat L, shown in plan at Figure 161. This is a brass plate 4 in. in diameter by 1/4 in. thick; the four water-ways m are 1/4 in. in diameter, and the centre hole N is drilled and tapped for a 1/4-in. stud O (Figure 159), on which the delivery valve works. The stud, after being screwed in, is riveted over on the under side as shown. The waterways are drilled with their centres 3/8 in. from the centre of the stud. The pulse-valve seat p is a brass plate 3 3/8 in. in diameter by 1/4 in. thick. A water-way 1 in. in diameter is cut in the centre, this being the diameter of the bore of the ram. Four holes for 1/4-in. bolts are drilled in both valve seats as shown. For larger rams the number of bolts will have to be increased, and should be spaced about 2 in. from centre to centre.

The delivery valve consists of a rubber disc Q and a brass disc R, both being 1 3/8 in. in diameter by 1/4 in. thick. Through the centre of each a hole is made for the stud O, which should fit loosely. The brass disc acts as a weight to keep the rubber disc in position. The nuts s are for adjusting the rise of the

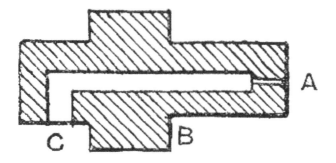

Figure 162 Section of Snifting Valve for Hydraulic Ram

valve. The delivery valve of small rams should not have a greater rise than 1/8 in., the rise being proportionately greater in the larger sizes. Through the stud o above the nuts a small hole is drilled for a small split pin, to prevent the nuts working loose.

The snifting valve, or air-inlet hole, is shown at t in Figures 159 and 160 (p. 145), a section being shown here at Figure 162. This consists of a brass plug with a 1/8-in. hole drilled through its centre, tapering to a very fine hole on the inner end A, where it should not be larger than would allow a very fine needle to pass through. The enlarged part B is cut hexagonal to take a small wrench, and at C a hole is drilled to meet the central hole, to guide the water downwards and prevent splashing, as a small quantity will squirt out on the closing of the pulse valve. The snifting valve is screwed into the T-piece through the lap of the flange K (see Figure 160) in place of one of the rivets. The object of this valve is to allow a little air to pass into the ram at each beat of the air vessel.

The details of the pulse valve are shown separately at Figures 163, 164, and 165. The spindle (Figure 163) of the valve is a 1/4-in. diameter mild steel rod 31/2 in. long with a thread cut at each end for a length of 11/4 in. A brass disc (Figure 164), 1 3/8 in. in diameter by 1/4 in. thick, tapered off at its outer edge as shown, forms the under support of a rubber disc, 1 3/8 in. in diameter by 1/4 in. thick. The upper brass disc (Figure 165) is

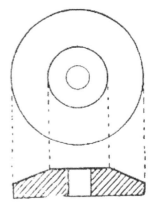

Figure 163 Details of Pulse Valve of Hydraulic Ram

Figure 164 Details of Pulse Valve of Hydraulic Ram

7/8 in. in diameter by 1/4 in. thick. The central hole in the brass disc is tapped to fit the thread of the valve spindle.

In putting the valve together the upper brass disc is first screwed up tight on the spindle with the flat side downwards (see Figure 159); over this is placed, the rubber disc; and over this is screwed the supporting brass disc, which must not be screwed up so tight as to press the rubber out of shape. Below this disc is a lock-nut; and further to prevent the disc and nut working loose while the ram is working, a small hole is drilled through the spindle for a small split pin.

At the other end of the spindle are two nuts and brass washers to adjust the length of the stroke and the weight of the pulse valve according to the length and fall of the drive pipe and the height the water is to be raised. The guide sleeve U for the pulse-valve spindle is made of a strip of sheet brass 3/4 in. wide by 3/16 in. or 1/4 in. thick, bent to the shape of a bridge as shown, the central part being 3/4 in. long and 1/2 in. in diameter; this sleeve must be fixed perfectly square with the lugs to allow the valve to rest square on its seating. The valves are faced with rubber in order to reduce noise and wear. In the case of the pulse valve, the rubber will assist the recoil or backward flow of the water immediately after the valve

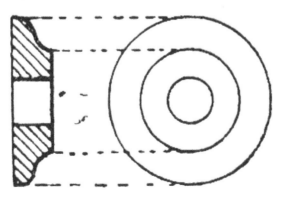

Figure 165 Detail of Pulse Valve of Hydraulic Ram

closes. This recoil is essential to the satisfactory working of a hydraulic ram.

The air vessel (see Figure 160) has an internal diameter of 2 1/4 in., and is made of sheet copper 1/16 in. thick, riveted together and soldered. Near the bottom is a short length of copper pipe V for the outlet to the discharge pipe, the diameter depending on the length of the delivery pipe; with a long delivery pipe the diameter should be not less than 1/2 in., but with a short delivery pipe 3/8 in. would be sufficient. The air vessel is fixed to the valve seating, as shown in Figure 159, by means of the flange W, which is riveted to the bottom of the air vessel. The width of the copper sheet to make the air vessel 2 1/4 in. in internal diameter, allowing 3/8 in. for the lap, is 83/4 in.; the length will depend on the capacity of the delivery pipe.

Figure 166 is a sectional elevation of the body in

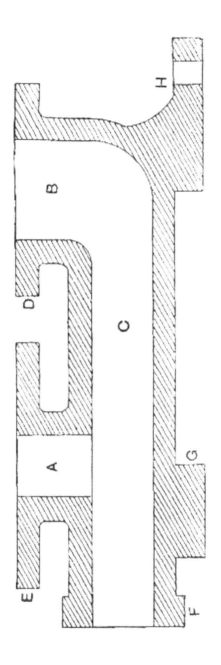

Figure 166 Section Hydraulic of Hydraulic Ram Body

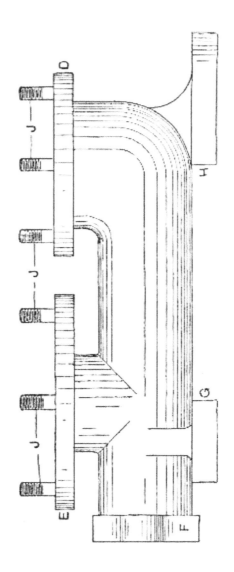

Figure 167 Side Elevation, Hydraulic of Hydraulic Ram Body

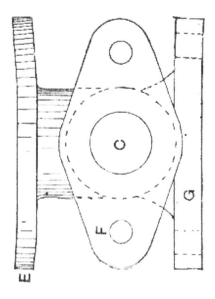

Figure 168 Side Elevation Hydraulic of Hydraulic Ram Body

cast-iron, Figure 167 a side elevation of the body, and Figure 168 an end elevation, showing the flange connecting to the drive pipe. The body is in one piece, and a pattern and core box will be required, the core box being for the waterway. The pulse and delivery valves, and the other fittings, are constructed in the same way as described for the copper ram; but for larger rams the pulse-valve guide may be an iron casting. In the illustrations a is the water-way to the delivery valve and air vessel; B, water-way to the pulse valve; C, waterway in the body; D, flange to which the pulse-valve seat and guide valve are bolted; E, flange to which are bolted the delivery valve seat and air vessel; F flange for connecting the ram to the drive pipe; G, back lugs for bolting the ram to the foundation, there being a bolt hole on each side of the ram (see Figure 168); H, a lug cast at the end of the elbow for bolting the front to the foundation; J, short studs, 1/4 in. in diameter, used for bolting the valve seats and the other fittings to the ram instead of bolts; the ram is held down to the foundation by three 3/8-in. holding-down bolts, two at the back of the ram and one at the front.

To obtain the highest degree of efficiency from a hydraulic ram water-raising plant, the horizontal length of the drive pipe to the ram should be the same length as the vertical height that the water is to be raised, and the diameter should be such as to discharge three times the available quantity. The diameter of the delivery pipe should be such as not to add more pressure on the ram than is due to a head of 3 ft. The area of the water-way of the pulse valve should be the same as the drive pipe, and the area of the delivery valve as large as possible, to allow of a large passage for the water with but a very small rise of the valve, thus preventing the water flowing back past the valve while closing, usually called " slip. " The capacity in cubic inches of the air vessel should not be less than the cubic capacity of the delivery pipe

When gauging streams with a very small quantity of water flowing, a triangular notched weir should he used, as this gives more accurate results with small quantities than a rectangular weir.

The formulae for making hydraulic ram calculations are: — where G = gallons of water flowing down the stream to work the ram; g = gallons of water raised; L = lift, or height the water is to be raised above the ram: f = the vertical fall of the feed water; and 0.65 the approximate efficiency of the ram described in this article.

To show the working of these formulae, a stream is taken as an example, having an available fall of 5 ft., with 2.819 gal. flowing per minute. It is required to raise the water to a tank 50 ft. above the ram, the length of the delivery pipe being 60 ft. What quantity of water would be raised to the tank in 24 hours? Applying Rule 1, the quantity raised, assuming that there is practically no frictional loss in the delivery pipe, will be $\frac{2.819 X 5}{50}$ X 0.65= .1832 gal. per minute, or .1832 x 60 x 24 = 263.8 gal., say 260 gal. in 24 hours, with 4,060 gal. passing through the ram in 24 hours. Now, applying Rule 2, that quantity of water will be required to raise .1832 gal. to a height of 50 ft. above the ram, the supply water having a fall of 5 ft. ? The quantity will be .1832 x 50 1 $\frac{.1832 X 50}{.5} X \frac{1}{0.65}$ = 2-819 gal. per minute. The working of these two examples will show the use of

$$(1) \quad g = \frac{G \times f}{L} \times 0{\cdot}65$$

$$(2) \quad G = \frac{g \times L}{f} \times \frac{1}{0{\cdot}65}$$

$$(3) \quad f = \frac{g \times L}{G} \times \frac{1}{0{\cdot}65}$$

the formulae, and from the data found by them the dimensions of any ram can be calculated.

When calculating the diameter of a ram, the simplest rules would be those of Box for finding the discharge and diameter of pipes. The rule for finding the diameter of a ram to pass a given quantity of water through a given length of drive pipe on a given fall, is the rule for finding the diameter of pipes. In applying this rule, the quantity that is to flow through the ram is multiplied by 3, the reason for this being that the water in the drive pipe only flows towards the ram about one-third of the time, and the diameter must be calculated for the maximum flow at any moment. Taking the above quantity of 2.819 gal. per minute on a fall of 5 ft., and assuming that the water is to be raised 50 ft. above the ram, the length of the drive pipe will be 50 ÷ 3 = 16 2/3 yd., say 17 yd. long, and

the maximum flow will be 2.819 x 3 = 8457 gal. The required diameter of the ram by this rule will be

$\sqrt[5]{8.457^2 x 17 / .5} \div 3 = 1$ in. in diameter

The maximum quantity that would pass through a ram of a given diameter, length of drive pipe, and fall is found by the rule for ascertaining the discharge of a pipe in gallons, and dividing the result by 3.

The diameter of the delivery pipe for a given frictional loss is found by the same rule as for finding the diameter of the ram; but instead of dividing by the available head, the head to be lost in friction is taken. In this case, the quantity flowing through the pipe is not multiplied by 3, as the water has a practically constant flow. With short delivery pipes, the head due to friction may be ignored, but for lengths above 50 yd. the head that will be lost must be added to the actual height the water is to be raised, and the diameter must be calculated for this loss.

The cubic capacity of the delivery pipe is found by multiplying the area of the pipe in square inches by

the length of the pipe in inches; the result will be the cubic capacity of the pipe in cubic inches, and the air vessel will have to be of the same cubic capacity. In the case of long delivery pipes, to prevent the air vessel being of inconvenient length it should be made balloon-shaped. By these simple rules it is possible to calculate the dimensions of a ram that would work satisfactorily with water from any stream.

Figure 169 shows the method of fixing the hydraulic ram, A being the vertical difference in feet between the water surface at the source of supply B and the ram C; D is the drive pipe, which is of the same length

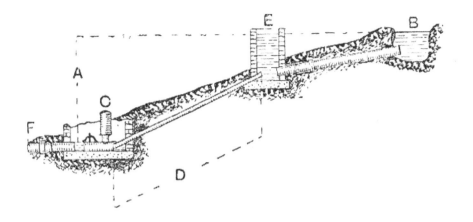

Figure 169 Method of Fixing Hydraulic Ram

as the vertical height to which the water is to be raised. As will be seen, the fall is an even gradient from the source of supply to the ram, and the distance between the ram and the source is greater than the length of the drive pipe; it will therefore be necessary to build the small brickwork tank E, the top of which must be a little higher than the surface of the water at the source. This tank is used to prevent the bursting of the pipes when the water is suddenly checked by the closing of the pulse valve, for with long pipes the pressure at the moment when the valve closes would be greater than the pipes and joints could withstand. The ram is bolted to a solid concrete foundation, and should be placed in a small house as shown. The level of the pulse valve must be at such a height that it is never covered by the tail-water, which is drained into a ditch at a lower level by the stoneware pipes F.

The drive pipe must be laid at an even gradient throughout. With small rams up to 2 in. in diameter this may be of wrought-iron with screw sockets, but for larger rams cast-iron flange pipes should be used, of a thickness sufficient to withstand the pressure without bursting. All the joints must be perfectly water-tight, and when wrought-iron pipes are used their ends must butt together in the sockets. To prevent the contraction of the column of water at its entry to the drive pipe, the end in the tank should be fitted with a trumpet mouth and flap valve, controlled by a chain, as shown in Figure 170.

The tank and source of supply are connected by ordinary stoneware socketed drain pipes of from twice to three times the diameter of the drive pipe, the joints being made with cement. To prevent debris from passing into the pipe, which after a time might become choked, a copper strainer should be fitted to the end (see Figure 171). The hole in the strainer should be of a greater area than the pipe. As a further protection against the strainer becoming blocked through weeds and rushes collecting round it, the strainer is surrounded by wire netting, secured to two posts driven into the bottom of the stream close to the bank, one post on each side of the pipe.

Figure 170 Flap Valve on Drive Pipe

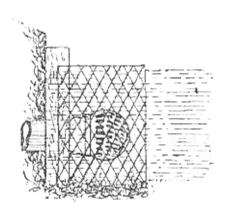

Figure 171 Detail of Strainer

The posts are driven in about 1 ft. away from the pipe on each side, and the netting fixed to allow a space of 1 ft. between the netting and the strainer.

The ram having been fixed in position and the drive pipe and the supply pipe to the tank laid, an attempt can be made to start the ram. The starting is effected by pressing down the pulse valve so as to let the water discharge and then allow it to rise, continuing to work it in this way for a few strokes by hand, when it should continue to beat by itself if the adjustment of the valve is suitable for the conditions of the water supply and the height the water is to be raised. But more probably it will be found that the ram will not work before it is adjusted.

The adjustment is effected by the two nuts and the brass washers at the end of the valve spindle. The nuts are for adjusting the length of the beat, and the washers for increasing or reducing the weight of the pulse valve by adding or taking away. The length of the beat and weight of the valve is controlled by the length of the drive pipe and the head of water above the ram, which works most satisfactorily when the stroke is regulated to be as long as possible, for with a short stroke it is more liable to stop.

On a slow and long stroke the ram will use and lift more water than on a fast and short one. The number of strokes is reduced and the length increased by unscrewing the nuts, and the number of strokes is increased and the length decreased by screwing them up. The weight of the pulse valve should be about the same as that of the water in the drive pipe when it is at rest. To adjust the valve to obtain the best results will occupy considerable time, as many trials will have to be made, by reducing or increasing, as required, the weight of the valve, and the number and length of the stroke, until a satisfactory result is obtained.

Stoppage of the ram after it has been working for a considerable time may be due to any of the following causes: The area of the drive pipe may be reduced or roughened by rust, or a deposit of organic matter may have formed on the inner surface of the pipe. In this case, although the head may be sufficient to dash the valve on to its seating, the force of the recoil may not be sufficient to allow the valve to open ready for another stroke. To remedy this, a bundle of wire should be dragged through the drive pipe several times, by means of a length of rope, after removing the ram from the pipe. Leaky joints in the drive pipe may also prevent the water recoiling sufficiently to open the pulse valve. The variation in level of the head water is a very common occurrence where the supply is intermittent. The level of supply being lowered, the air is carried in with the water, causing the valve to chatter for a short time, and ultimately to remain closed, though if the valve is a heavy one it will remain open. The air vessel may be water-logged, owing to the shifting valve becoming inoperative, through being covered by the tail-water; or the shifting valve may be choked, causing the resistance in the air vessel to be so great that the delivery valve will not open, thus stopping the delivery of water, though the pulse valve will continue to beat, and everything will appear to be in perfect working order. The recoil of the drive water may be insufficient to allow the pulse valve to open. With a slow lift, the absence of air in the air vessel may sometimes be detected by the water being delivered in a succession of spurts accompanied by concussion in the pipes. To remedy this, shut the flap valve on the end of the drive pipe in the tank, remove and empty the air vessel, and see that the shifting valve is in good order. On replacing the air vessel, the ram should again deliver water properly.

To stop the ram, the pulse valve is held up for a few seconds against its seating, when it will cease to beat; but to stop the ram for examination and repairs, the head or flap valve in the tank must be closed.

###

ABOUT THE PUBLISHER

This book has been published by EnergyBook - RW Jemmett

EnergyBook.co.uk

DISCLAIMER

The information in this publication has been supplied in all good faith and believed to be correct. However no liability will be accepted for any accident, damage or injury caused as a result, or arising from the use of information from this publication. Information contained in this work has been obtained by EnergyBook from sources that are believed to be reliable. However, neither EnergyBook or its authors or editors guarantee the accuracy or completeness of any information published herein. Please be careful

Made in the USA
Las Vegas, NV
17 June 2025